European Women in Chemistry

Edited by
Jan Apotheker and
Livia Simon Sarkadi

Related Titles

Garcia-Martinez, J., Serrano-Torregrosa, E. (eds.)

The Chemical Element

Chemistry's Contribution to Our Global Future

2011

ISBN: 978-3-527-32880-2

Ravina, E.

The Evolution of Drug Discovery

From Traditional Medicines to Modern Drugs

2011

ISBN: 978-3-527-32669-3

Ghosh, A.

Letters to a Young Chemist

2011

ISBN: 978-0-470-39043-6

Armaroli, N., Balzani, V.

Energy for a Sustainable World

From the Oil Age to a Sun-Powered Future

2010

ISBN: 978-3-527-32540-5

Rehder, D.

Chemistry in Space

From Interstellar Matter to the Origin of Life

2010

ISBN: 978-3-527-32689-1

European Women in Chemistry

Edited by
Jan Apotheker and Livia Simon Sarkadi

WILEY-VCH

WILEY-VCH Verlag GmbH & Co. KGaA

The Editors

Dr. Jan Apotheker
University of Groningen
Department for Chemistry Education
Nijenborgh 4
9747 AG Groningen
The Netherlands

Dr.habil. Livia Simon Sarkadi
Budapest University of Food Technology
and Economics
Department of Applied Biotechnology and
Food Science
Muegyetem rkp. 3
1111 Budapest
Hungary

Cover

The cover idea and material was kindly
provided by Rita Tömösközi Farkas.

Library of Congress Card No.:
applied for

British Library Cataloguing-in-Publication Data
A catalogue record for this book is available from the
British Library.

**Bibliographic information published by the Deutsche
Nationalbibliothek**
The Deutsche Nationalbibliothek lists this publication in the Deutsche Nationalbibliografie; detailed
bibliographic data are available on the Internet at <
HYPERLINK "http://dnb.d-nb.de" http://dnb.d-
nb.de>.

Composition Typodesign Hecker, Leimen
Printing and Binding Strauss GmbH, Mörlenbach
Cover Design Adam Design, Weinheim

Printed in the Federal Republic of Germany
Printed on acid-free paper

ISBN: 978-3-527-32956-4

Foreword

"A book about Women in Chemistry, what a strange project: how could so few women bring something to chemistry?" I anticipate that this will not be an uncommon reaction to the publication of the book "European Women in Chemistry". It is true that there are not many world-famous women chemists. To look at the place given to women in science, let us have a look at Nobel laureates, who are among the most prominent scientists: between 1901 and 2010, the Nobel Prizes for Sciences and the Prize in Economic Sciences were awarded to 612 laureates, of which 17 were women. And if we now consider the chemistry Nobel laureates, the Nobel Prize in Chemistry has been awarded to 159 laureates, among which 4 were women (1911, Marie Curie, field of nuclear chemistry, *"in recognition of her services to the advancement of chemistry by the discovery of the elements radium and polonium, by the isolation of radium and the study of the nature and compounds of this remarkable element"*; 1935, Irène Joliot-Curie, field of nuclear chemistry, *"in recognition of their synthesis of new radioactive elements"*; 1964, Dorothy Crowfoot Hodgkin, field of biochemistry, structural chemistry *"for her determinations by X-ray techniques of the structures of important biochemical substances"*; 2009, Ada Yonath, biochemistry, structural chemistry, *"for studies of the structure and function of the ribosome"*.

Why so few? First, because people were convinced that Science was rigorous and rational and women were supposed to be weak and irrational. As a consequence, women scientists have been systematically excluded from doing serious science; they generally encountered their family's – mostly father's – resistance to their studying. *"apprenez-leur qu'il doit y avoir, pour leur sexe, une pudeur sur la science presqu'aussi délicate que celle qu'inspire l'horreur du vice"* (tell them that their sex must have for science as much a sense of decency as that inspired by the horror of vice) (Fénelon, traité de l'éducation des filles, 1687). Furthermore, as women were excluded from the high schools that prepared men for university, if they wanted to learn science, they had to hire tutors. This explains why the few scientifically educated women were, for a long time, encountered mainly in the rich and intellectual classes of society.

Anyway, as far as chemistry is concerned, men can do chemistry, but women do the cooking. With regard to chemistry-like activities performed by women, they were often associated with perfumes, ointments, poisons and, as a consequence, with witchcraft. Consequently, we can assume that many women who knew the

European Women in Chemistry. Edited by Jan Apotheker and Livia Simon Sarkadi
Copyright © 2011 WILEY-VCH Verlag GmbH & Co. KGaA, Weinheim
ISBN 978-3-527-32956-4

properties of plants (the first natural product chemists), were often victims of obscurantism and burned as witches...

A look at the destiny of women chemists shows that their lives were seldom plain ones, and that most of them had difficult or extraordinary fates. This is probably one of the reasons for the great influence these women had, and still have, for example as models for young people – and not only girls. Indeed, it is much more exciting to try to identify oneself with an out-of-the-ordinary-person, than with one having an uneventful story; and as, at least some years ago, most women chemists had uncommon stories, it is not surprising that they are considered by students as better models than male chemists. They campaigned for more vocational opportunities such as the right to vote and a state-supported secondary and higher education for girls. They certainly succeeded in the latter cause and, thanks to their struggle and determination, by the beginning of the 20th century women in several countries were finally accepted into Universities. Now, even if some discrimination against women in science still exists, women chemists must cope with this and understand that their future depends more on what they want to do themselves than on what others want. By doing this, they will show once more their determination and how strong-willed they can be.

Nicole Moreau
Charenton, France

Contents

European Women in Chemistry. Edited by Jan Apotheker and Livia Simon Sarkadi
Copyright © 2011 WILEY-VCH Verlag GmbH & Co. KGaA, Weinheim
ISBN 978-3-527-32956-4

Preface

One of the reasons for 2011 being chosen as the International Year of Chemistry is the centennial commemoration of the Nobel Prize awarded to Maria Skłodowska-Curie. This centenary led to the idea of a book to show the range of female chemists active across Europe in what many would suggest is still a male-dominated profession.

The chapters cover women from alchemical times up to the 19 and 20th centuries when women gained access to higher education. The individual subjects were suggested by EuCheMS member societies and a final decision was taken by the editors; as in any such selection there are other subjects who might have been included. Indeed it is hoped that the book will initiate discussion and debate about this.

The stories demonstrate both the range of activities of female chemists and just how difficult it was for them, and female scientists in general, to develop rewarding careers. Unfortunately, in most European countries this situation only began to change after 1960. Until this date the vast majority of women chemists experienced great problems in securing an academic career despite their excellent quality.

In this book we have focused on academic careers. Other careers of women that have a chemical background have not been included. Otherwise Margaret Thatcher, Angela Merkel and other politicians with a chemical background would certainly have been included.

Nowadays there are several scholarship pograms to encourage female scientists, both at European and national levels. There are also networks for women scientists to share experiences and offer support to students and young scientists starting out on their career paths.

The editors hope that you will enjoy reading the different stories about female chemists from different countries, with different backgrounds. It is not intended to be a book to finish in one reading, rather it is a book to inspire young women to consider a career in chemistry. It should, however, not only be read by women; male chemists should ask themselves how their careers would have developed had they been faced with the same obstacles. Teachers of chemistry in secondary and tertiary education would also benefit from reading this book so that they can ensure that the opportunities for a career in science are not inadvertently directed at their male students.

European Women in Chemistry. Edited by Jan Apotheker and Livia Simon Sarkadi
Copyright © 2011 WILEY-VCH Verlag GmbH & Co. KGaA, Weinheim
ISBN 978-3-527-32956-4

We would like to acknowledge the various people from Wiley who helped so much in getting this book together. The EuCheMS Presidency, who initially suggested a book on this topic, and all the authors who contributed to this book, are to be thanked. Without their support, encouragement and enthusiasm the project would not have been possible. Particular thanks are due to Professor Nicole Moreau (President of IUPAC) who has written a foreword to the book.

Jan Apotheker
Livia Simon Sarkadi

About the Editors

Jan Apotheker is a lecturer in Chemistry Education at the University of Groningen. After obtaining his academic degrees from the University of Groningen in Biochemistry, he taught chemistry at a local secondary school for 25 years. One of his prime responsibilities as lecturer is the training of teachers in all levels of education. He is also involved in the organization of outreach activities both from the university and on a national scale. He is a member of the steering committee 'New Chemistry' that is currently developing a new chemistry curriculum for secondary education in the Netherlands. Jan is the Royal Dutch Chemical Society board member for education, an IUPAC Committee Member for chemistry education, and a member of the EUCHEMS division for chemistry education.

Livia Simon Sarkadi is a Professor of Applied Biotechnology and Food Science at the Budapest University of Technology and Economics, Hungary. Since 1980, she has taught biochemistry, food chemistry, and food analysis. She has supervised a number of PhD, BSc and MSc students. Besides being an author and co-author of many scientific papers, she wrote a textbook on Biochemistry. She is a member of the Editorial Board of International Journals (European Food Research and Technology, Food and Nutrition Research). She has been the Chair of the Food Protein Working Group of the Hungarian Academy of Sciences since 1996 and is currently the Chair of the EuCheMS Food Chemistry Division, and an elected member of the EuCheMS Executive Board.

European Women in Chemistry. Edited by Jan Apotheker and Livia Simon Sarkadi
Copyright © 2011 WILEY-VCH Verlag GmbH & Co. KGaA, Weinheim
ISBN 978-3-527-32956-4

List of Contributors

Jean-Pierre Adloff
Société chimique de France
250, Rue Saint-Jacques
75014 Paris
France

Katharina Al-Shamery
University of Oldenburg
Fak. V, IRAC
Postfach 2503
26111 Oldenburg
Germany

Didier Astruc
Société chimique de France
250, Rue Saint-Jacques
75014 Paris
France

Susanne Bartel
University of Oldenburg
Fak. V, IRAC
Postfach 2503
26111 Oldenburg
Germany

Christiane Bonnelle
Société chimique de France
250, Rue Saint-Jacques
75014 Paris
France

C.W. Mineke Bosch
University of Groningen
Faculty of Arts, Modern Hist.
Postbus 716
9700 AS Groningen
The Netherlands

Marco Ciardi
University of Bologna
Department of Philosophy
Via Zamboni 38
40126 Bologna
Italy

Danielle M.E. Fauque
Société chimique de France
250, Rue Saint-Jacques
75014 Paris
France

Miriam Focaccia
University of Bologna
Department of Philosophy
Via Zamboni 38
40126 Bologna
Italy

Carl G. Gahmberg
University of Helsinki
Dept. of Chemistry
A.I. Virtasen aukio 1
00014 Helsinki
Finland

European Women in Chemistry. Edited by Jan Apotheker and Livia Simon Sarkadi
Copyright © 2011 WILEY-VCH Verlag GmbH & Co. KGaA, Weinheim
ISBN 978-3-527-32956-4

Jean-Pierre Genet
Société chimique de France
250, Rue Saint-Jacques
75014 Paris
France

Sally Horrocks
University of Leicester
School of Historical Studies
University Road
Leicester, LE1 7RH
United Kingdom

Henryk Kozlowski
University of Wroclaw
Faculty of Chemistry
F. Joliot-Curie 14
50-383 Wroclaw
Poland

Katalin Nyári-Varga
Hungarian Museum for Science
and Technology Budapest
Kaposvár u. 13 -15
1117 Budapest
Hungary

Marianne Offereins
Jodichemdreef 40
3984 JT ODIJK
The Netherlands

István Próder
Hungarian Museum for Science
and Technology Budapest
Kaposvár u. 13 -15
1117 Budapest
Hungary

Pekka Pyykko
University of Helsinki
Dept. of Chemistry
A.I. Virtasen aukio 1
00014 Helsinki
Finland

Maria Rentetzi
MPI for the History of
Science
Boltzmannstr. 22
14195 Berlin
Germany

Renate Strohmeier
Uni-Klinik Frankfurt
Gynäkologie und Geburtshilfe
Theodor-Stern-Kai 7
60590 Frankfurt am Main
Germany

Brigitte van Tiggelen
Voie du Vieux Quartier 18
1348 Louvain-la-Neuve
Belgium

Éva Vámos
Hungarian Museum for Science
and Technology Budapest
Kaposvár u. 13 -15
1117 Budapest
Hungary

Annette B. Vogt
MPI für Wissenschafts-
geschichte
Boltzmannstr. 22
14195 Berlin
Germany

Maria the Jewess

Marianne Offereins

Maria the Jewess was an alchemist who probably lived in Alexandria, Egypt, in the first or the third century. Although no facts are known about her life, there are many references to Maria in ancient texts. Because alchemy was a secretive science, perhaps to protect its practitioners from persecution, it was not uncommon for alchemists to write under the name of a deity or a famous person. Maria wrote under the name of Miriam the Prophetess, sister of Moses.

Fragments of her work, including one called the *Maria Practica*, are extant in ancient alchemical collections. She also may have been the author of *The Letter of the Crown and the Nature of the Creation by Mary the Copt of Egypt* which was found in a volume of Arabic alchemical manuscripts, translated from the Greek. In this work the major theories of Alexandrian alchemy are summarized and several chemical processes described including the manufacture of colored glass. Maria was often quoted by other early alchemists, particularly the Egyptian encyclopedist and alchemist Zosimos of Panopolis (third or fourth century), the alchemist and writer Olympiodoros (fifth or sixth century) and Michael Maier (seventeenth century). Zosimos states that Maria was the first to prepare copper burnt with sulfur, the 'raw

Maria the Jewess

European Women in Chemistry. Edited by Jan Apotheker and Livia Simon Sarkadi
Copyright © 2011 WILEY-VCH Verlag GmbH & Co. KGaA, Weinheim
ISBN 978-3-527-32956-4

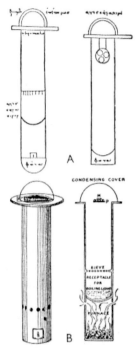

Kerotakis (www.alchemywebsite.com).

material' for the preparation of gold. She taught that the 'Great Work' could only be prepared in the early spring and that God had given its secret exclusively to the Hebrews. Maria believed that all matter is basically one, and that success in making gold will come when parts are joined: "One becomes two, two becomes three, and by means of the third the fourth achieves unity, thus two are but one". In her writings there is an analogy between humankind and the metals: "Join the male and the female, and you will find that which is sought after".

Her theoretical contributions remained influential into the Middle Ages and beyond, but Maria was even more famous for her designs of laboratory apparatus. Maria invented, and improved on, techniques and tools that remain basic to laboratory science today and in her writings she described her designs for laboratory apparatus in great detail. Distillation was essential to experimental alchemy, so Maria invented a still or *alembic* and a three-armed still called the *tribikos*. The liquid to be distilled was heated in an earthenware vessel on a furnace. The vapor condensed in the *ambix*, which was cooled with sponges, and a rim on the inside of the *ambix* collected the distillate and carried it to three copper delivery spouts fitted with receiving vessels.

For her experiments she invented the *kerotakis*, her most important contribution to alchemical science: a cylinder or sphere with a hemispherical cover, placed on a fire. Suspended from the cover at the top of the cylinder was a triangular palette,

Balneum mariae (www.alchemywebsite.com).

used by artists to heat their mixtures of pigment and wax, and containing a copper–lead alloy or some other metal. Solutions of sulfur, mercury, or arsenic sulfide were heated in a pan near the bottom of the cylinder. The sulfur or mercury vapors condensed in the cover and the liquid condensate flowed back down, attacking the metal to yield a black sulfide called 'Mary's Black'. This was believed to be the first step of transmutation. A sieve separated impurities from the black sulfide and continuous refluxing produced a gold-like alloy. Plant oils such as attar of roses were also extracted using the *kerotakis*.

Her water bath, the *balneum mariae*, was similar to a double-boiler and was used to maintain a constant temperature, or to slowly heat a substance. Two thousand years later, the water bath remains an essential component of the laboratory. One should not confuse the *balneum mariae*, where the inner vessel is heated with steam to get a temperature above 100 °C and the *'bain marie'* in which the temperature remains under 100 °C.

Maria the Jewess was one of the first chemists to combine the theories of alchemical science with the practical chemistry of the craft traditions, and, therefore, can be considered as one of the founders of western chemistry.

Literature

Alic, M. (1986) *Hypatia's Heritage. A History of Women in Science from Antiquity to the Late Nineteenth Century*, The Women's Press, London.

Kass-Simon, G. (1993) *Women of Science. Righting the Record*, Indiana University Press, Bloomington and Indianapolis.

Lennep, J. van (1984) *Alchemie*, Gemeentekrediet België, Brussels.

Ogilvie, M. (2000). *The Biographical Dictionary of Women in Science. Pioneering Lives from Ancient Times to the Mid-20th Century*, Vol 2, Routledge, London and New York.

Cleopatra the Alchemist

Marianne Offereins and Renate Strohmeier

■ Like Maria the Jewess, Cleopatra the alchemist, also known as Cleopatra the Gold-maker, probably lived in the third century and is associated with the school of Maria the Jewess.

Like Maria the Jewess 'Cleopatra' is most probably a pseudonym.

Cleopatra was a philosopher and a practical experimentalist and is often confused with Cleopatra the physician, who lived at approximately the same time and who is mentioned in the work of Hippocrates.

What remains of Cleopatra's work are a discourse and a single surviving papyrus sheet with symbols and diagrams. A copy is in the library of the University of Leiden, the Netherlands. In the discourse, which is written as a dialogue, she compares the philosopher-alchemist who contemplates his work to a loving mother who thinks about her child and feeds it. According to Lindsay in his book *The Origins of Alchemy in Graeco-Roman Egypt* this discourse was "the most imaginative and deeply-felt document left by the alchemists".

The papyrus, the Chrysopoeia (Gold-making), pictures the archetypical symbol of the Ouroboros, a serpent eating its tail (symbol of infinity), and a double ring on which is the inscription: "One is the Serpent which has its poison according to two compositions, and One is All and through it is All, and by it is All, and if you have not All, All is Nothing".

Within the ring are the symbols for gold, silver and mercury. In other parts of the papyrus are a dibikos (a two-armed still) and a kerotakis-like apparatus. The drawings on the right-hand side could be representing the transformation of lead into silver.

Cleopatra investigated weights and measures, attempting to quantify the experimental side of alchemy. Her texts were used until the late Middle Ages, with many alchemists referring to her work.

Like Maria she also used the sun and dung as laboratory heat sources. So, if we are busy developing ways to use the sun and dung as energy sources, we have very important predecessors.

European Women in Chemistry. Edited by Jan Apotheker and Livia Simon Sarkadi
Copyright © 2011 WILEY-VCH Verlag GmbH & Co. KGaA, Weinheim
ISBN 978-3-527-32956-4

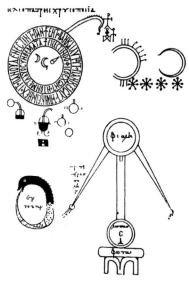

The chrysopeia (gold-making) of Cleopatra
(http://library.du.ac.in/xmlui/bitstream/handle/1/788/
Ch8%20Alchemy.pdf?sequence=14).

Literature

Alic, M. (1986) *Hypatia's Heritage. A History of Women in Science from Antiquity to the Late Nineteenth Century*, The Women's Press, London.

Kass-Simon, G. (1993) *Women of Science. Righting the Record*, Indiana University Press, Bloomington and Indianapolis.

Lennep, J. van (1984) *Alchemie*, Gemeentekrediet België, Brussels.

Lindsay, J. (1970) *The Origins of Alchemy in Graeco-Roman Egypt*, Muller, London.

Rebière, A. (1897) *Les Femmes dans la Science, Notes Recueillies*, Librairie Nony & Cie, Paris.

Strohmeier, R. (1998) *Lexicon der Naturforscherinnen und Naturkundigen Frauen Europas. Von der Antike bis zum 20. Jahrhundert*, Harri Deutsch Verlag, Thun und Frankfurt am Main.

Perenelle

Marianne Offereins

■ The date of birth of Perenelle (1320 (or 1340) – 1402 (1412)) and her origins are still not very well known. She lived in fourteenth century Paris, where she – after being widowed twice – married Nicolas Flamel, a well-to-do scribe in 1355. They lived in the Rue des Écrivains, near the church Saint-Jaques-de-la-Boucherie.

They have become famous through the books of J.K. Rowling and Michael Scott, where they are described as alchemists who found the philosophers stone and consequently found the source of eternal life.

In 1357 Flamel bought for two florins the manuscript that would change their lives. Flamel writes about it: "(...) a gilded Book, very old and large. It was not of Paper, nor of Parchment, as other Books be, but was only made of delicate rinds (as it seemed unto me) of tender young trees. The cover of it was of brass, well bound, all engraved with letters, or strange figures; and for my part I think they might well be Greek Characters, or some-such-like ancient language. Sure I am, I could not read them, and I know well they were not notes nor letters of the Latin nor of the Gaul for of them we understand a little. As for that which was within it, the leaves of bark or rind were engraved, and written with admirable diligence, with a point of Iron, in fair and neat Latin letters, coloured. It contained thrice-seven leaves, for so were they counted in the top of the leaves, and always every seventh leaf was without any writing; but, instead thereof, upon the first seventh leaf, there was painted a Rod and Serpents swallowing it up".

The title was written in big gilded letters: *Abraham Eleazar le Juif, prince lévite, astrologue et philosophe, à la gent des Juifs par l'ire de Dieu dispersé aux Gaules, Salut.*

During the next twenty-one years Flamel and Perenelle worked on the translation of the book, which should contain the secret of the transmutation and the philosophers stone. They consulted many people and attempted many experiments themselves – but to no avail. Finally, Flamel travelled to Spain, where he met a Jewish physician who explained to him the meaning of the text and the figures. After that they worked for three years and, finally, on Monday 17 January, 1382 Flamel wrote in his book, *Livre des Figures*, that Perenelle and he transformed half a pound of mercury into 'pure silver'. And on April 25 they made from "the red

European Women in Chemistry. Edited by Jan Apotheker and Livia Simon Sarkadi
Copyright © 2011 WILEY-VCH Verlag GmbH & Co. KGaA, Weinheim
ISBN 978-3-527-32956-4

(a)

(b)

(c)

(a) Portal of the Cimitière des Innocents, Uit.
(b) P. Arnauld (1612) Le Livre des figures hiéroglifiques, Paris.
(c) Woodcut of Nicolas Flamel and his wife Perenelle.

stone" "almost as much pure gold". As a commemoration he had "(...) painted in the fourth Arch of the Church-yard of the Innocents, as you enter in by the great gate in St. Dennis-street, and taking the way on the right hand, the most true and essential marks of the Art, yet under veils, and Hieroglyphical covertures, in imitation of those which are in the gilded Book of Abraham the Jew (...)".

Perenelle died in 1397 on September 11, and left her husband a fortune of 5300 pounds.

The problem with Flamel and Perenelle is that there are no contemporary sources. The earliest sources date from the sixteenth century. It is said that because they found the philosophers stone they are still living...

Literature

http://www.levity.com/alchemy/testment.htm l (accessed 24 December 2009).

Alic, M. (1986) *Hypatia's Heritage. A History of Women in Science from Antiquity to the Late Nineteenth Century*, The Women's Press, London.

Federmann, R. (1964) *Die Königliche Kunst. Eine Geschichte der Alchemie*, Paul Neff, Wien.

Lennep, J. van (1984) *Alchemie. Bijdrage Tot de Geschiedenis van de Alchemistische Kunst*, Gemeentekrediet België, Brussels.

Rebière, A. (1897) *Les Femmes dans la Science. Notes Recueillies*, Librairie Nony & Cie, Paris.

Anna, Princess of Denmark and Norway, Electress of Saxony (1532–1585)

Renate Strohmeier

■ Owner of the largest and finest chemical laboratory established in sixteenth century Germany, Anna is one of the few female chemists/alchemists of the sixteenth century of whom we know. Since she has some importance in the history of Saxony, the historians of the nineteenth century wrote her biography and evaluated her extensive correspondence. In her letters she describes her interests and activities, often to other women who were engaged in the same field of knowledge. There is not much available data on other sixteenth century women alchemists, like Isabella Cortese (?–1561) or Marie Meurdrac (seventeenth century?), whom we only know of because they published treatises on chemistry. The occult science of alchemy was dangerous – even life threatening – and could carry women quickly to the stake.

In the sixteenth century, when chemistry was rather alchemy, Paracelsus (1493–1541) established medical treatment with chemical substances based on the healing power of plants and minerals. Newly awakened scientific curiosity in combination with astrology, Hermetic ideas and traditional superstitious beliefs, led to the development of iatrochemistry (pharmacy), the field of Anna's scientific activities. A lot of new laboratory equipment and procedures were invented in the early sixteenth century, and she applied these in her laboratories. Most important of these were the improved distillation apparatus for her well known Aqua vitae.

In Annaburg, Saxony, a town that was named after her, she established a kind of "plant site" for the production of pharmaceuticals. The 200 square steps facility with walls and moats, sheltered distillation houses and laboratories of amazing size. One of the houses was as big as a church, had self-supporting vaults and many chimneys. A visitor reports: *"he saw a laboratory with sixteen chimneys which contained furnaces in the shape and height of horses, lions and apes and one in the shape of an eagle with outspread gold-plated wings"*. In these laboratories alls kinds of ingredients were processed into medical products. Herbal ingredients came from her own gardens or were collected in the woods and fields of the neighborhood by lo-

European Women in Chemistry. Edited by Jan Apotheker and Livia Simon Sarkadi
Copyright © 2011 WILEY-VCH Verlag GmbH & Co. KGaA, Weinheim
ISBN 978-3-527-32956-4

Anna, Princess of Denmark and Norway, Electress of Saxony.

cal herb-collecting women. Large amounts of leaves, fruits, roots and flowers were dried and stored. However, not only plants, but also remedies from the animal kingdom, such as pulverized human leg bones, moss grown on human skulls, human lard, ox bile, dog fat, horse and donkey milk, deer and goat blood, and, not to be forgotten, the highly coveted Unicorn, were mixed into ointments, syrups, electuaries and medical aquavits. After her death 181 ingredients for her healing waters were found in the storerooms and laboratories of Annaburg. These remedies seem rather quaint today, however, these drug components are described in many dispensaries of the sixteenth century.

As Anna knew no Latin, one can assume that she received no higher education. Her knowledge and lively interest in medicine and its production was probably awakened in her childhood by her mother, because later on it became the main subject of the correspondence between mother and daughter. Her early teacher of the art of Aqua vitae distillation was countess Anna von Mannsfeld. Advanced contemporary knowledge and the new procedures of her time mostly came from the medical attendants of the court. Dr. Paul Luther (1533–1593) a doctor and alchemist, may have been her most significant teacher. Letters of inquiry to all important doctors and alchemists of their time were found in Anna's and August's correspondence. For example they asked doctor Ch. Pithopoeus to teach them "the foundations of his new science and medicine, which effects by extraction of the main powers and things (active substances) in the fire". Learned doctors were not the only source of her medical knowledge. Anna collected formulas of all kinds of contemporary healers like herbal women, quack doctors, shepherds and barbers.

The castle of Annaburg, built by Anna and August I of Saxony (1572–1575).

Her large collection of recipes and medical cures were ordered and supplemented by pharmacists and doctors in her pharmacopoeia.

Together with her husband she was also engaged in alchemistic experiments. With the help of the Swiss chemist Sebald they created "three ounces of gold out of six ounces of silver within six days" in 1578. In 1585 they gave some "acranum, made by their own hands", to the count of Brandenburg, who gratefully accepted the "lapidi de rebus". These activities were rather dangerous for women of her time. Anna's high social status as a Princess may have saved her from being suspected of witchcraft and being sentenced to the stake.

Literature

Carl von Weber (1865) *Anna, Churfürstin von Sachsen*, Tauchniz, Leipzig

Harless, J.C.F. (1830) *Die Verdienste der Frauen um Naturwissenschaft, Gesundheits- und Heilkunde, so wie auch um Laender-, Voelker- und Menschenkunde, von der aeltesten Zeit bis auf die neueste : ein Beitrag zur Geschichte und geistiger Cultur, und der Natur- und Heilkunde insbesondere*, Vandenhoeck-Rupprecht, Goettingen

Keller, K. (2007) Anna von Dänemark, in *Sächsische Biografie*, ed. Institut für Sächsische Geschichte und Volkskunde e.V., revised by Martina Schattkowsky, Online: http://www.isgv.de/saebi/

Marie Meurdrac (1600s)

Marianne Offereins und Renate Strohmeier

■ Author of one of the first treatises on chemistry by a woman.

Biographical data about Marie Meurdrac's life are difficult to obtain. Proof of her existence is her treatise on chemistry, which was first published in 1666 in Paris. *La Chymie Charitable et Facile, en Faveur des Dames* is considered the first treatise on chemistry by a woman since the works of Maria the Jewess about 1600 years earlier. Marie Meurdrac may have known of this early colleague of hers because she writes concerning the *Bain-marie Distillation*: "This Distillation is called by the name of the woman who invented it, who was the sister of Moses, Marie, called the Prophetess, who wrote the Book entitled *The Three Words*".

Marie Meurdrac describes the content of her book as follows: "I have divided this Book into Six Parts: in the first, I treat principles and operations, vessels, lutes, furnaces, fires, characteristics and weights: in the second, I speak of the properties of simples (medical herbs or medicines made from such plants), of their preparation and of the method of extracting their salts, tinctures, fluid and essences: the third treats Animals, the fourth Metals: the fifth treats the method of making compound

Sixteenth century chemistry laboratory, engraving after a drawing of Pieter Bruegel the Elder, 1560.

European Women in Chemistry. Edited by Jan Apotheker and Livia Simon Sarkadi
Copyright © 2011 WILEY-VCH Verlag GmbH & Co. KGaA, Weinheim
ISBN 978-3-527-32956-4

medicines, with several tested remedies: the sixth is for Ladies, in which there is a discussion of everything capable of preserving and increasing beauty. I have done my best to explain myself well and to facilitate the operations: I have been very careful not to go beyond my knowledge, and I can assure that everything I teach is true, and that all my remedies have been tested; for which I praise and glorify God". (Translation by Bishop and DeLoach, 1970).

The book contains a table of 106 alchemistical symbols and a table of weights used in medicine. According to the alchemical tradition she assumed that substances were formed on three principles: salt, sulfur and mercury. Some passages of the book suggest that she was not only an alchemist/chemist but also a medical doctor. She assures for instance "I have used it (essence of rosemary) with good results and have affected some admirable cures with it".

In her introduction Marie Meurdrac describes the "inner struggle" between the traditional concept of a woman, which she claimed "remain silent, listen and learn, without displaying ... knowledge" and "...on the other hand, I flattered myself that I am not the first lady to have had something published". She describes her motivation to "let the book leave my hands"... "that it would be a sin against Charity to hide the knowledge that God has given me, which may be of benefit to the whole world".

Her anticipation that the book would not achieve success because "men always scorn and blame the products of a woman's wit" did not come true. It had two more French editions (1680 and 1711) and was translated into German (editions in 1673, 1676, 1689 and 1712) and Italian.

Literature

Bishop, L.O. and DeLoach, W.S. (1970) Marie Meurdrac – First Lady of Chemistry? *J. Chem. Educ.*, **47** (6), 448–449.

Meurdrac, M. (1680) *La Chymie Charitable et Facile, en Faveur des Dames*, 2nd ed., Chez Jean Baptiste Deville, Lyon.

Tosi, L. (2001) Marie Meurdrac: Paracelsian chemist and feminist. *Ambix*, **48** (2), 69–82.

Emilie Le Tonnelier de Breteuil, Marquise du Châtelet (1706–1749)

Marianne Offereins

She was one of the most famous Femmes savantes and had a great influence on Voltaire and his work. Because of her translation of Newton's *Principia Mathematica* into French and the addition of her own commentary, her influence on the introduction of the ideas of Newton in France was great.

Gabrielle Emilie Le Tonnelier de Breteuil was born in Paris in 1706. Her father, Nicolas Breteuil Le Tonnelier baron de Preuilly, was chief of protocol at the royal court, where in his youth he had caused quite a lot of scandals. When he was 45 years old, he married Gabrielle Anne de Froulay, about whom not much more is known other than that she came from the higher nobility and was educated in a

Portrait of la marquise du Châtelet (1740) by Nicolas de Largillière (1656–1746), documentation du Louvre.

European Women in Chemistry. Edited by Jan Apotheker and Livia Simon Sarkadi
Copyright © 2011 WILEY-VCH Verlag GmbH & Co. KGaA, Weinheim
ISBN 978-3-527-32956-4

convent. The education they gave to their children, consisted mainly of advice such as: "Blow your nose in your napkin" and "Never comb your hair in church".

As a child Emilie impressed her father enough with her intelligence to convince him that some education would not be wasted on her. Moreover, because she did not meet the beauty standards of her time – she was tall for her age and was said to have "a skin as a grater" – so as a "born old maid" she needed a good education. From the time she was about six years old, she was taken into the care of the best available governesses and teachers. She had a natural sense of language and soon mastered English, Latin and Italian. She studied Milton, Virgil, and Tasso, and translated the Aeneid. At the age of 19 she married the 34 years old Marquis du Châtelet. As he was a colonel in the Guard regiment, he often was away from home for a long time. During his absence Emilie did not have time to be bored, she amused herself with a series of lovers.

About her appearance, opinions varied: the ladies found her ugly, the men thought her extremely attractive.

In the first two years of their marriage the couple had two children, a girl and a boy. When Emilie was 27 years old, another boy was born. After his birth she began, on the advice of the Duc de Richelieu (grandnephew of the Cardinal), to seriously study mathematics and natural philosophy. Neither her husband nor her children could prevent her from having a busy social life at court as well, where she moved in the intimate circle of the Queen.

Here she made two 'unforgivable errors': she refused to finish her study, which for a woman was regarded as highly inappropriate; and, even worse, in the spring of 1733 she started a relationship with Voltaire, who would remain her regular companion for the rest of her life, even when, later, both had fallen in love with someone else. As for Voltaire, after the publication of his *Lettres Philosophiques*[1] (also called *Lettres Anglaises*), Paris became increasingly dangerous for him, therefore Emily persuaded her husband to provide shelter for Voltaire on their estate at Cirey sur Blaise in Lorraine, at a safe distance from the court. Together they took care of the restoration of the dilapidated castle. There was an extensive library and a fully equipped laboratory, with ovens, air pumps, a telescope, and a microscope, where Emilie could work on her experiments. Here, she was visited by the important scholars of her time, including Pierre-Louis Moreau de Maupertuis, one of the leading mathematicians and astronomers of his time, his pupil, mathematician Johann Samuel König, Alexis Claude Clairault, and the Bernoulli brothers. The contact with these scientists was so important to Emilie that she dressed as a man in order to be admitted into the coffee-houses where the men had their discussions.

Emilie was strongly influenced by Maupertuis, who accompanied her in her studies. König also helped her for a short time with her studies, but after a difference of opinion that collaboration ended.

Life on Cirey was definitely not exclusively devoted to study. Because Voltaire was a lover of theater, 'la belle Emilie' regularly organized entire theater performances.

---

[1] In this book he announced the rationalist ideas of the Enlightenment.

She studied very much, it was said of her that she did not need more than one or two hours sleep each night and that she was spectacularly healthy.

Her first publication, *Sur la Nature du Feu* (1738), she wrote because she disagreed with Voltaire on the subject. She wrote this work at night in secret. Whenever she felt she was sleepy, she dipped her hands into ice water to stay awake.

From the moment Voltaire could show himself again in Paris, Emilie and he divided their time between Paris and Cirey.

Both Voltaire and Maupertuis were great admirers of Newton's ideas and eager to spread 'Newtonian' ideas in France. Maupertuis made the ideas of Newton a fashionable topic for the Salons, and Voltaire encouraged Emilie du Châtelet to translate Newton's work. This time Emilie wrote *Institutions de Physique* (1740), for use in the education of her son. The usual books on physics education were now about 80 years old and Emilie wanted a book in which the modern ideas of Leibnitz and Newton were mentioned. Samuel König took his revenge by telling everyone in Paris that the work was simply a repetition of his lectures. After that she translated the *Principia Mathematica* by Newton and added her own algebraic comments. There is no doubt that these books influenced Voltaire and therefore Emilie can be placed among known scholars such as Clairault, the Bernoullis, Mairan and Maupertuis.

In 1748 Emilie began a relationship with the Marquis of Saint-Lambert, a courtier and second-rate poet. When she discovered she was pregnant by her lover, Voltaire helped her organize a visit by her husband to Cirey. Three weeks later he left, believing that he would be a father again. The baby was born in early September 1749. Voltaire wrote that the girl was born while her mother worked at her desk on her notes on Newton. The newborn baby was placed on a geometry book, while Emilie placed her papers together and was taken to bed. For a few days everything went well until suddenly Emilie died, probably of puerperal fever or, as other sources say, of a pulmonary embolism, a few days later she was followed by her daughter.

In France Emilie du Châtelet is best known for the letters she left and for her *Discours sur le Bonheur.*

Her intelligence and character are undisputed.

Frederick II of Prussia wrote about her to Voltaire: "That Emilie reminds me, is very flattering to me. Be so kind as to assure her that I have very high opinion of her, for Europe she belongs to the great men"(!).

Literature

Alic, M. (1986) *Hypatia's Heritage, a History of Women in Science from Antiquity to the Late Nineteenth Century*, The Women's Press, London.

Ehrman, E. (1986) *Mme Du Châtelet, Scientist, Philosopher and Feminist of the Enlightenment*, Berg Publishers, Oxford.

Mozans, H.J. (1974) *Woman in Science*, with an introductory chapter on woman's long struggle for things of the mind, facsimile of the 1913 edn, MIT Press, Cambridge, MA.

Mozans, H.J. (1913/1991) *Women in Science*, University of Notre Dame Press, New York, Notre Dame, Indiana/London.

Ogilvie, M.B. and Harvey, J. (eds) (2000) *The Biographical Dictionary of Women in Science. Pioneering Lives from Ancient Times to the Mid-20th Century*. Routledge, Cambridge, MA/London.

Osen, L.M. (1974) *Women in Mathematics*, The MIT Press, Cambridge, MA.

Phillips, P. (1990) *The Scientific Lady, a Social History of Woman's Scientific Interests 1520–1918*, Weidenfeld and Nicholson, London.

Schiebinger, L. (1989) *The Mind Has No Sex?* Harvard University Press, Cambridge, MA.

Marie Lavoisier (1758–1836)

Marianne Offereins

For many people, Lavoisier's Law will be familiar. However, there will be fewer who know that Antoine Lavoisier was helped by his wife Marie in doing his experiments. She made a significant contribution to her husband's work.

On January 20, 1758, Marie Anne Pierette Paulze was born in Montbrison, in the province Loire, in France. Her father, Jacques Paulze, worked primarily as a parliamentary lawyer and financier. Most of his income, however, came from running the Ferme Générale (The General Farm) which was a private consortium of financiers who paid the French monarchy for the privilege of collecting taxes. Marie had two brothers, and when she was three years old her mother died. Marie

Marie Lavoisier and her husband,
by Jacques Louis David (1788).

European Women in Chemistry. Edited by Jan Apotheker and Livia Simon Sarkadi
Copyright © 2011 WILEY-VCH Verlag GmbH & Co. KGaA, Weinheim
ISBN 978-3-527-32956-4

proved to be a smart girl, who was educated in a convent, as befitted a French girl of her social class.

When she was thirteen, the Count d'Amerval proposed to marry Marie but, as he was nearly three times her age, her father tried to object to the marriage. This proved to be rather difficult and he was threatened with losing his job with the Ferme Générale. Therefore he proposed to his colleague Antoine Lavoisier that he should ask for his daughter's hand instead. Lavoisier, a French nobleman, who had already achieved fame as a chemist and had been elected to the Academy of Sciences in 1768, accepted the proposition, and he and Marie-Anne were married on 16 December 1771. By that time Lavoisier was already about 28 years old.

Marie soon became interested in his scientific research and began to actively participate in his laboratory work. Antoine continued her education, but now the lessons centered around the use of balances, burning lenses, and reduction vessels, and German and Latin, the languages of the scientific community. To help her husband with his investigation of the physical nature of fire and heat, she taught herself English, so she could translate the American and British articles he needed into French. She also took art lessons from the French painter Jacques-Louis David, the one who painted the famous portrait of the Lavoisier couple, and began illustrating Antoine's articles.

The Lavoisiers spent most of their time together in the laboratory, working as a team conducting research on many fronts. In fact, most of the research in the laboratory was actually a joint effort between Antoine and Marie. She helped him with his experiments, made all the notes, kept the laboratory reports and carried out their scientific correspondence. Especially, Marie's particular drawing gift came in handy, because she made sketches of the experiments and the experimental tools. Lavoisier's treatise *Traité élémentaire de Chimie* (1789), which must be regarded as the first modern chemistry text, in which he describes 23 elements which are the basis of all chemical reactions, contains engravings in her hand.

Another major contribution to science was made by her translation of the works of English authors into French. She translated the chemical treatises of Henry Cavendish, Joseph Priestley and other important British scientific researchers. Her translation of *Essay on Phlogiston* by Richard Kirwan, with comments by Lavoisier and his colleagues, was of the utmost importance: the until then widely held theories of combustion, which maintained that the element phlogiston was essential to combustion, proved untenable. In their experiments the Lavoisier couple showed that Phlogiston did not exist.

Most important to science, Antoine formulated the law of conservation of matter, which established that there is no gain or loss of weight in the elements of a chemical reaction, a theory that bound chemistry to physical and mathematical laws. As a team, they established modern chemistry by separating its scientific aspects from alchemy and by evolving an updated scientific glossary. They coined the term "oxygen", identified it as an elemental gas, described the oxidation process that changes iron to rust, and analyzed the products of normal human respiration as water and carbon dioxide.

During the early years of their marriage their home became a gathering place for members of the French intellectual community.

When the fury of the Revolution overtook the country, the position of Lavoisier, who like Marie's father was a member of the *Ferme Générale,* was particularly vulnerable. Pretty soon he was arrested and put in prison, in addition, all his possessions were confiscated. During his captivity, Marie worked tirelessly, but unsuccessfully, for his release. On May 8, 1794, at the end of Robespierre's 'Reign of Terror', Antoine Lavoisier was guillotined (as well as Marie's father and many of their friends). Marie was arrested too, based on certain incriminating documents, but she was released after 65 days in the Bastille. She came out penniless as a consequence of the confiscation of her land. She had to take refuge in the care of a former servant. About a year later, most of Lavoisier's possessions were returned to her. Most important for science was the confiscated scientific library, which she intended to keep for the future. In 1792 Lavoisier had started to make detailed notes of his experiments for publication. At the time of his death only a part of it was ready. Marie finished his work, and in 1805 she published the *Mémoires de Chimie,* (Memoirs of Chemistry) under the name of her deceased husband. She published the work in two volumes along with her original introduction. She distributed free copies to known French scientists.

During the Directory, and later under the reign of Napoleon, as matters became less violent in Paris, again she could welcome visitors to her salon. Several well

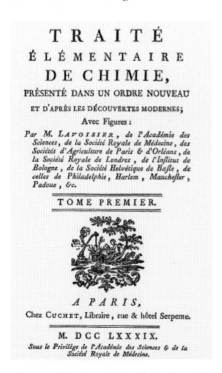

Title page of the first volume of *Mémoires de Chimie*.

known scientists courted her, one of her suitors was chemical magnate Pierre Samuel Dupont de Nemours, but she preferred the American physicist, Benjamin Thompson, better known as Count Rumford of Bavaria, founder of the Royal Institution of Great Britain, whom she married in 1805 after a four year courtship. After her marriage, she insisted on being called Countess Lavoisier-Rumford. The marriage was not a success and after four years it ended in divorce. After her divorce from Rumford she worked as a successful businesswoman and she was also known for her charitable work. As the years went by, continuing her chemist's work became increasingly difficult for her, but for many years she received in her salon well known scientists, including Cuvier, Berthollet, Humboldt and others. She died in Paris at the age of 77.

As Marie Lavoisier's scientific work was so closely intertwined with the work of her husband, it is difficult to specify exactly what can be attributed to her. Together they brought a fundamental change, replacing the mysterious practices of the alchemists with systematic chemical principles.

Through her drawings, translations, explanations of notes, and arranging the publication of Lavoisier's 'Memoirs of Chemistry', she made an important contribution to scientific knowledge.

Literature

Alic, M. (1986) *Hypatia's Heritage, a History of Women in Science from Antiquity to the Late Nineteenth Century*, The Women's Press, London.

Offereins, M. I.C. (1996) *Vrouwen Miniaturen uit de exacte vakken*, VeEX, Utrecht.

Ogilvie, M. and Harvey, J. (eds) (2000) *The Biographical Dictionary of Women in Science. Pioneering Lives from Ancient Times to the Mid-20th Century*, Routledge, Cambridge MA / London.

Schiebinger, L. (1991) *The Mind Has No Sex? Women in the Origins of Modern Science.* Harvard University Press, Cambridge MA / London.

Thijsse, W.H. (1985) *Rokoko, Democratie in Wording*, De Walburg Pers., Zutphen.

http://www.answers.com/topic/marie-paulze-lavoisier

Jane Haldimand Marcet (1769–1858)

Marianne Offereins

Jane Marcet wrote one of the most popular books on chemistry. For almost a century her *Conversations on Chemistry* was the most used book everywhere in Europe and in America.

Jane Haldimand was the only girl of the twelve children of Anthony Francis Haldimand, a wealthy Swiss merchant, who lived in London. In her childhood she often visited her relatives in Geneva, Switzerland. From the age of 15, after the death of her mother, Jane took care of the household and of her younger brothers. As a child Jane was educated by tutors together with her brothers in her father's home. Subjects were, as in all well-to-do families, natural philosophy (science) as well as languages and history. Here she showed a special interest in art and botany. After her marriage in 1799 to Dr. Alexander Marcet (1770–1822), another London-based Swiss, who had graduated from medical school at the University of Edinburgh in 1797, but who preferred to spend his time as an amateur chemist, Jane al-

Jane Haldimand Marcet (http://www.rsc.org/images/
FEATURE-marcet-300_tcm18-87786.jpg).

European Women in Chemistry. Edited by Jan Apotheker and Livia Simon Sarkadi
Copyright © 2011 WILEY-VCH Verlag GmbH & Co. KGaA, Weinheim
ISBN 978-3-527-32956-4

so took an interest in chemistry. The couple eventually had three children, their son François was to become a distinguished physicist, not much is known of the other children.

After the death of father Haldimand, Jane Marcet inherited enough money for her husband to stop working as a physician and to concentrate entirely on his real interest – chemistry. As Alexander Marcet was a fellow of the Royal Society the Marcets frequently attended Sir Humphry Davy's entertaining demonstrations on chemistry at the Royal Institution, but Jane often found the science confusing. To better understand these lectures, Jane Marcet attended other lessons at the Royal Institution. Luckily, her husband was very adept in clarifying the concepts for her, and Jane became convinced that this conversational style of teaching was highly effective. Somewhat curiously she concluded that this was especially so for the female sex, "whose education is seldom calculated to prepare their minds for abstract ideas, or scientific language". The Marcet couple moved in a circle of prominent intellectuals, including historian Henry Hallam, political economists Thomas Malthus and Harriet Martineau, novelist Maria Edgeworth and naturalists Augustin-Pyramus de Candolle, Auguste de la Rive and mathematician and astronomer Mary Somerville. Jane became involved in the activities of this group and, encouraged by her husband, she started her own writing career.

She wrote a number of introductory science books, especially intended for women and young people. In her introduction she wrote: "In venturing to offer to the public, and more particularly to the female sex, an Introduction to Chemistry, the author, herself a woman, conceives that some explanation may be required; and she feels it the more necessary to apologize for the present undertaking, as her knowledge of the subject is but recent, and as she can have no real claims to the title of chemist".

Although – as she assured her readers – she neither pretended to be a scientist nor sought a depth of knowledge that might be "considered by some (...) as unsuited to the ordinary pursuits of her sex", she did believe that "the general opinion no longer excludes women from an acquaintance with the elements of science"(*Conversations on Chemistry* III).

Her first book, *Conversations on Chemistry*, was published in 1806. Following the publication of her enormously successful first book she wrote *Conversations on Political Economy*, which was highly praised and with which she acquired a celebrity 'similar to that of a man'. Encouraged by this success, Jane wrote *Conversations on Natural Philosophy*. Now she felt on thin ice! As she mentions in the preface, she was not sufficiently literate in mathematics and physics to achieve the required level. Therefore, this book was meant for very young children.

Some textbooks in the first half of the nineteenth century laid an emphasis on the spiritual content: the acquisition of knowledge to obtain more admiration for the Creation. Other books treated domestic issues like the rising of dough, keeping milk and butter and the properties of fuels, for women, while textbooks for things like soil analysis, tanneries and medication were provided for men.

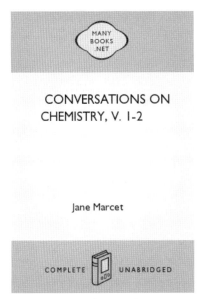

Conversations on Chemistry by Jane Marcet (taken from:
http://manybooks.net/titles/marcetj2690826908-8.html#).
The book may be downloaded from this site.

The chemistry method of Jane Marcet, however, is both theoretical and practical, and provides an insight into the 'real chemical' experiments, such as the production of nitrous oxide (N_2O) by gentle heating of ammonium nitrate.

'Conversations on Chemistry' consists of 26 lessons, or 'conversations'. The material is systematically built using the latest insights. Each lesson is as follows: a beautiful, elegant lady, "Mrs B.", is teaching two young girls. One, Emily, is an inquisitive, intelligent girl, about 12 years old; the other, Caroline, around 13 years old, the daughter of the manager of a lead mine, had no interest at all in chemistry. Emily, asks smart questions, while Caroline is very critical and more interested in explosions than in basic science. Marcet explains in a foreword: "otherwise the book would be too boring". Because of the many experiments, with clear drawings, the practice is an integral part of the theory. As the source of heat An oil lamp is used as the source of heat, which provides enough heat for the usually mild reactions. Gases are collected and stored in a pig's bladder.

Mrs B. encourages the girls to use a language which is not too scientific. "You can better say 'rust' instead of 'oxidation', else other people might find you a drama queen"!

Marcet's book was an immediate success. In the same year, 1806, as the first edition was released in England, an edition appeared in America. From 1806 to 1850 there were 23 printings, sometimes several editions in a single year. A contemporary estimate is that in America about 160 000 copies were sold. Marcet's Conversations is not intended as a textbook, and in England the book was used as it was intended, a guide to the then popular lectures on chemistry or science. But in America it became the most successful basic chemistry method of the first half of the nineteenth century. A large quantity of – male – publishers adapted the book for classroom use, and generally the work is also attributed to these publishers. Copyright did not exist in America at that time, so Jane Marcet did not have any influence over this and of course did not receive any income from it.

Michael Faraday got hold of the Conversations on Chemistry in 1810 when he was an apprentice at bookbinder Riebau. Later they became good friends, and Jane always included his new work, as well as Davy's, in each edition. By this text began his love of chemistry. Later, after he had performed experiments, Michael Faraday wrote : "I felt that I had got hold of an Anchor of chemical knowledge, and clung fast to it".

The lively discussions indeed stimulate one to read and experiment. In technical training institutions and medical schools for young men *Conversations on Chemistry* was the textbook which was used as the first introduction to chemistry. Over the years many parts of the book have become available on the internet. In many cases even for free, so everybody can see how useful the book still is.

Literature

Alic, M. (1986) *Hypatia's Heritage, a History of Women in Science from Antiquity to the Late Nineteenth Century*, The Women's Press, London.

Clarke, J. (1984) *In our Grandmothers' Footsteps*, Virago Press, London.

Mozans, H.J. (1913/1991) *Women in Science*, University of Notre Dame Press, New York.

Ogilvie, M. and Harvey, J. (eds) (2000) *The Biographical Dictionary of Women in Science. Pioneering Lives from Ancient Times to the Mid-20th Century*, Routledge, Cambridge MA/London.

http://www.jstor.org/pss/4028037 (accessed 25-2-2010).

http://www.rsc.org/chemistryworld/restricted/2007/June/ThewomanthatinspiredFaraday.asp (accessed 25-2-2010).

http://www.gutenberg.org/files/26908/26908-h/Conver1.html (accessed 26-2-2010).

http://www.ncbi.nlm.nih.gov/pmc/articles/PMC1033865/pdf/medhist00141–0081.pdf (accessed 25-2-2010).

Julia Lermontova (1846–1919)

Marianne Offereins

■ Julia Lermontova was the first woman in the world to obtain a degree in chemistry. Her contemporaries regarded her as one of the most important chemists of her time. She only worked as a chemist until the age of 35. All her life Lermontova stood in the shadow of her friend Sofia Kovalevskaja, the mathematician who became the first female Professor in Europe.

On December 21 in 1846 (according to the Julian calendar) or on January 2, in 1847 (according to the Gregorian calendar), Julia Vsevolodovna was born into the aristocratic Lermontov family in St. Petersburg. She was the daughter of Elisawjeta Andrejevna Kossikovsky and her husband General Vsevolod Lermontov, who was a second cousin of the famous Russian poet Mikhail Lermontov. Julia was raised in the Greek Orthodox tradition as well as in the Roman Catholic tradition. During her youth she lived in Moscow where her father was in charge of the Moscow Cadet Corps.

Her parents belonged to the Moscow intelligentsia and gave the education and training of their children a high priority. That is why at the Lermontov residence there were often different foreign governesses and nannies at the same time. For the children only the best private teachers were good enough.

Although the family could not follow Julia in her interest for science, they did not prevent the development of her knowledge in this area. She read the necessary professional literature, and conducted simple experiments at home.

Initially Julia wanted to study medicine but both the sight of skeletons in the dissecting room, and the poverty of the patients were so repulsive to her, that she decided to sign in at the Petrovskaia Agricultural College (now 'Timirjasew-College') in Moscow, which had an excellent chemistry program. Although her application was supported by a large number of professors, she was rejected. So she decided to go abroad. It sounds easy, but for women, and especially for Russian women, at the time it certainly was not an easy task, needing a great deal of courage, perseverance and a strong personality. The study was difficult and the majority of women had little financial resources: the money was for study by the boys and the men. In addition, there was often the opposition of men.

European Women in Chemistry. Edited by Jan Apotheker and Livia Simon Sarkadi
Copyright © 2011 WILEY-VCH Verlag GmbH & Co. KGaA, Weinheim
ISBN 978-3-527-32956-4

Lermontova and Kovalevskaya
(http://www.serednikovo.ru/history/lermontovy/
lermontova_j_v/lermontova_j_v.html)

Through her cousin Anna Evreinova, later the first female doctor in law, she met Sofia Korwin-Krukowskaja who entered into a convenience marriage with Vladimir Kovalevski to be able to study abroad as a married woman. Sofia Kovalevskaia persuaded Julia's parents to let their daughter go. After all, if she was in the company of a married woman, she had her chaperone.

In the autumn of 1869 Julia arrived in Heidelberg, where she stayed with the Kovalevskis. As a result of the energy of Kovalevskaia, Julia was admitted into the laboratory of Bunsen, who was known as a woman-hater.

> *"Er[Bunsen] habe sich verschworen, kein Frauenzimmer, namentlich keine Russin in sein Laboratorium aufzunehmen. Er habe also auch Frl. Lermontof nicht bei sich wollen arbeiten oder hören lassen. Da seiest Du [Kovalevskaja] zu ihm gekommen und habest ihn so allerliebst gebeten, daß er nicht habe widerstehen können und seinen Vorsätzen ungetreu geworden sei. Ist dem wirklich so? Dann würde der Philosoph doch nicht ganz Unrecht gehabt haben. Bunsen steht allerdings in dem Ruf, daß er bei seinen Erzählungen ein wenig phantasiert. Dichtet er auch doch Romane, wenn er sie auch nicht publicirt."*
> Karl Weierstraß in a letter addressed to Sofia Kovalevskaja (1874)

Weierstraß, also wrote:

*"Es war dann natürlich viel von Dir und Deinen beiden Studien-
genossinnen [Lermontova und Jewreinova] die Rede, und ich
habe manches aus eurem damaligen 'Zigeunerleben' erfahren,
das mich sehr amüsiert hat. Ihr seid Gegenstand sehr großer
Aufmerksamkeit bei den Heidelbergern gewesen."*

At Bunsen's laboratory she researched into platinum compounds. The two Russian women were soon joined by Anna Jevreinova whose parents disapproved strongly of foreign study. Her father would 'rather have had her dead than at a University'. As a convenience marriage proved impossible, she fled over the border, under fire from the guards.

In 1871 Lermontova followed her friend Kovalevskaya from Heidelberg to Berlin. There Lermontova worked in the private laboratory of August Wilhelm Hofmann and as a private student she attended his lectures on Organic Chemistry. She published her first research: '*Ueber die Zusammensetzung des Diphenins*'.

At the beginning of 1874, she finished her doctoral dissertation: 'Zur Kenntniss der Methylenverbindungen', and after a lengthy discussion on the admissibility of women, and more specifically 'Julie von Lermontoff', she could defend her work in a regular graduation ceremony, 24 October 1874 in Göttingen, where she was preceded by Dorothea Schlözer, the first woman who attained a doctor's degree in Göttingen.

She had been very anxious for her graduation and it was therefore a pleasant surprise for her to see the professors sitting at a tea table with pastry, and even wine. The exam was not easy, but afterwards they drank and ate, and, moreover, the professors told her that the doctors title was granted in the first degree: *magna cum laude*.

When in 1874 Julia returned to Russia, Dmitri Mendeleev and the other chemists of the Russian Chemical Association were very happy to see her. For a short time she worked in the laboratory of Vladimir Markovnikov in Moscow, but after some time she went back to St. Petersburg, where she found a job as a coworker with Alexander Butlerov and M. L'vov in the Laboratory of the University. Among others she did research on the production of 2-methyl-2-butenoic acid. From 1876 she worked as a correspondent for the French '*Bulletin de la Société Chimique de Paris*' .

In the same year Julia was infected with typhoid fever. As a complication she got severe encephalitis. People feared for her life as well as for her intelligence, but fortunately she fully recovered.

In 1877 Julia's father died. Therefore, she moved to Moscow to attend to her family affairs. In Moscow she found a place in the Laboratory of Markovnikov who worked on oil research, because oil was found in large quantities in the vicinity of Baku. She was also the first woman to work in this area of research. She developed a device for continuous distillation of petroleum which was highly praised by her contemporaries although it was not possible to use the device on an industrial scale.

From St. Petersburg, Butlerov still tried to convince her to accept the position of instructor at the Superior courses for women, but she did not accept the offer. According to Julia herself this was because she wondered whether she would be given permission by the Minister of Education, but according to Butlerov komma Kovalevskaya was to blame, because she had left her daughter fully in the care of Julia, while Sofia, according to Butlerov, "tramped around the world".

In 1881 Julia, became the first woman member of the Russian Technical Association.

Because Julia had inherited the family estate Semenkovo, she had made a habit of living there for a number of months in the summer. In the end this won favor over chemistry and she settled permanently in the country. There she shifted her interest to the agricultural sciences. The cheese that was developed on the estate was a success and was sold all over Russia and the Ukraine. On the estate Julia led a retired life. However, it is well known that in the spring of 1889 she became seriously ill with double pneumonia, and in the autumn of that year she travelled to Stockholm to visit Kovalevskaya. In May 1890 she met Sofia Kovalevskaya in St. Petersburg and at the same time she picked up Fufa, Sofia's daughter. The sudden death of Kovalevskaya in 1891 moved her deeply and she wrote her 'memories of Sofja Kovalevskaya'.

After the October revolution in 1917 an attempt was made to nationalize the estate Semenkovo, but Anatoli Lunascharski, the Minister of Education who also played a part in the protection of the heritage of Ekaterina Gontscharova, intervened, and Julia was allowed to keep the estate. In December 1919, Julia Lermontova died from a brain haemorrhage. Although Julia had never married, her stepdaughter 'Fufa' Kovalevskaya saw her as her own 'Mama Lulia'. She inherited the whole Lermontova estate.

Literature

Koblitz, A.H. (1983/1993) *A Convergence of Lives. Sofia Kovalevskaia: Scientist, Writer Revolutionary*, Rutgers University Press, New Brunswick NJ.

Ogilvie, M. and Harvey, J. (eds) (2000) *The Biographical Dictionary of Women in Science. Pioneering Lives from Ancient Times to the Mid-20th Century*, Routledge, Cambridge, MA/London.

Rebière, A. (1897) *Les Femmes dans La Science. Notes Recueillies*, Librairie Nony & Cie, Paris.

Rogger, F. (1999) *Der Doktorhut im Besenschrank. Das abenteurliche Leben der ersten Studentinnen – am Beispiel der Universität Bern*, eFeF Verlag, Bern.

Roussanova, E. (2001) *Julia Lermontova (1846–1919), Die erste promovierte Chemikerin des 19. Jahrhunderts*, Hamburg

Tobies, R. (1997) *Aller Männerkultur zum Trotz. Frauen in Mathematik und Naturwissenschaften*, Frankfurt am Main

Martha Annie Whiteley (1866–1956)

Sally Horrocks

■ In 1903 Martha Annie Whiteley became the first female member of staff at the Royal College of Science, from 1907 part of Imperial College. She was active in the battle to secure admission for women to the Chemical Society and was the first woman to be elected to the Society's Council, serving from 1928 to 1931. At Imperial College, where she eventually attained the rank of Assistant Professor, she managed to combine an active research career with significant contributions to undergraduate teaching, and played a major role in the Imperial College Women's Association that she founded in 1912. During World War One she worked on a range of government projects. In 1920 she was appointed to the Order of the British Empire (OBE) for this work. She was a longstanding editor of *Thorpe's Dictionary of Applied Chemistry*, continuing in this role, initially alongside Jocelyn Field Thorpe, after her official retirement in 1934.

Martha Annie Whiteley was born in 1866 in London, the second daughter of William Sedgewick Whiteley, a house agent's clerk and his wife Hannah Bargh. She attended Kensington Girl's School and then Royal Holloway College, where she gained a University of London BSc in chemistry in 1890. The following year she passed the Oxford honours mathematical moderations and joined the staff of Wimbledon High School as a science teacher. In 1898 Whiteley started studying part-time at the Royal College of Science (RCS) and in 1900 moved to the post of science lecturer at St Gabriel's Training College in Camberwell. Her research at the RCS on the organic chemistry of barbiturate compounds earned her a University of London DSc in 1902. She was invited to join the staff of the RCS in 1904 as a teaching scholar, turning down a post at Royal Holloway College to do so. In 1905 she was promoted to assistant and then to demonstrator in 1908. She was awarded a four-year British Federation of University Women fellowship in 1912, promoted to lecturer in 1914 and Assistant Professor in 1920, two years after she attained the Fellowship of the Royal Institute of Chemistry. Whiteley formally retired in 1934 but continued to work as an editor and contributor to *Thorpe's Dictionary of Applied Chemistry*. She was the principal editor of the twelve volumes of the enlarged fourth edition after her co-editor, Jocelyn Field Thorpe, died in 1939. She was 88

European Women in Chemistry. Edited by Jan Apotheker and Livia Simon Sarkadi
Copyright © 2011 WILEY-VCH Verlag GmbH & Co. KGaA, Weinheim
ISBN 978-3-527-32956-4

Martha Annie Whiteley (By permission of the College Archives, Imperial College London).

years old when this project was finally completed in 1954. It has been suggested that Whiteley resisted promotion to the rank of full professor because she saw it as an impediment to continuing as an active researcher, but her publication record seems limited when compared with those of her male professorial colleagues. This could be attributed to the delayed start to her academic career, to her reluctance to include her name on every published paper to which she had contributed, or her extensive involvement with *Thorpe's Dictionary*.

Another explanation for her limited number of publications might be the additional pastoral responsibilities that she took on within Imperial College and her commitment to expanding women's opportunities in science. She was an active member of the British Federation of University Women and founded the Imperial College Women's Association in 1912. Two years earlier she had successfully lobbied for improved cloakroom facilities for women staff and students. This was an early example of the way that she actively worked to encourage women students not only in her own department but throughout the College, where she is said to have been known as the 'Queen Bee'. Beyond Imperial College she was active in the long campaign to persuade the Council of the Chemical Society to admit women to the Fellowship of the Society. This was successful in 1920 and along with fellow campaigner Ida Smedley Maclean she founded the Women's Dining Club of the Chemical Society. In 1928 she was the first woman elected to serve on the Council of the Society. Imperial College honored her with a Fellowship in 1945 in recognition of her services to chemical science and contributions to the College.

Little is known of Whiteley's personal life and like most other women of her generation who had successful careers, she never married. Her Chemical Society obituarist A. A. Eldridge listed her recreations as 'domestic and social duties' and in her retirement she seems to have continued to devote herself to editing and other chemical matters rather than developing other interests. She retained strong friendships with her former colleagues and students, had close links to Imperial College and strong religious convictions.

Whiteley's first published scientific work was a joint publication with Karl Pearson in the *Proceedings of the Royal Society of London*, 1899, 'Data for the problem of evolution in man. I. A first study of the variability and correlation of the hand'. She appears to have undertaken many of the measurements and the 'laborious arithmetical reductions'. This work was probably carried out before she embarked on her research at the RCS. Her initial chemical research was on the organic chemistry of barbiturate compounds and the tautomerism in oximes, particularly mesoxamide and related compounds. This was her focus from 1898 until the demands of World War One diverted her attentions to the synthesis of drugs and the improvement of their production processes. Her first publication on chemical research, 'The oxime of mesoxamide and some allied compounds', appeared in 1900 in the *Journal of the Chemical Society, Transactions*. Her wartime work involved the production of hydrochlorine and lactate for β-eucaine and diethylamionoethanol for novocaine and the production of sugars. She also developed two compounds for use on the battlefield, a tear-gas known as SK (ethyl iodoacetate) and an incendiary mixture that was named DW (Dr Whiteley) in her honour. After the war she continued to publish on the oxime of mesoxamide and on malonyl derivatives and received funding from the Royal Society to support her research. Among her coauthors during this period were two other female chemists, Dorothy Yapp ('The reaction between diaxonium salts and malonyldiurethane', *J. Chem. Soc.* (1927), 521–528) and Edith Hilda Usherwood ('The oxime of mesoxamide (isonitrosomalonamide) and some allied compounds. Part III Ring formation in the tetra-substituted series, *J. Chem Soc. Trans.* (1923) 123, 1069–1089). Usherwood later married Christopher Ingold who was a member of staff in the Chemistry Department at Imperial College from 1920 to 1924.

During the 1920s Whiteley contributed to *Thorpe's Dictionary of Applied Chemistry* and was coauthor, with her Imperial College colleague Sir Jocelyn Field Thorpe, of *A Student's Manual of Organic Chemical Analysis* (1925). Her collaboration with Jocelyn Thorpe continued after her formal retirement in 1934, first as co-editor of a supplement to *Thorpe's Dictionary* (1936) and then as compilers of an entirely new edition that started to appear in 1941.

Martha Whiteley was exceptional among her generation of women chemists in that she managed to sustain a long and successful career as an academic in a major British institution of higher education at a time when very few women were able to do this. After a delayed start, during which she earned a living as a schoolteacher, her career more closely resembled that of those men who might be regarded as in the second rank of academic chemists rather than those of other women who continued as researchers but did not obtain secure positions or work independently from male mentors. In common with male colleagues who did not obtain chairs or Fellowships of the Royal Society, she made significant contributions to the teaching of students, published original research regularly but not voluminously and made notable contributions during World War One that were rewarded by the state. She also served the chemical community through her roles in the Chemical Society and as indefatigable editor of a major reference work and endeavored to inspire other women to pursue science and to make it easier for them to do so. Her calm efficiency ensured that she was able to operate effectively in what had been exclusively a man's world and demonstrate through her presence that women, when given the opportunity, could forge a successful and productive career in chemistry.

Literature

Barrett, A. Whiteley, Martha Annie (1866–1956), in *Oxford Dictionary of National Biography*, online edn, Oxford University Press, Sept. 2004. http://www.oxforddnb.com/view/article/ 46421 (30 July 2010).

Creese, M.R.S. (1997) Martha Annie Whiteley (1866–1956), chemist and editor. *Bulletin for the History of Chemistry*, **20**, 42–45.

Creese, M.R.S. (1991) British women of the nineteenth and early twentieth centuries who contributed to research in the chemical sciences. *British Journal for the History of Science*, **24**, 275–305.

Eldridge, A.A. (1957) Martha Annie Whiteley, 1866–1956, *Proceedings of the Chemical Society* **1**, 182–183.

Gay, H. (2007) *The History of Imperial College London 1907–2007: Higher education and research in science, technology and medicine*, Imperial College Press, London.

Imperial College Centenary Website, http://www.imperial.ac.uk/centenary/ default.shtml (accessed 28 July 2010)

Mason, J. (1991) A forty years' war, *Chemistry in Britain*, **27**, 233–238.

Owen, L.N. (1956) Dr M. A. Whiteley OBE, *Nature*, **177**, 1202–1203.

Rayner-Canham, M and Rayner-Canham, G. (2008) *Chemistry Was Their Life: Pioneer British Women Chemists, 1880–1949*, Imperial College Press, London.

Agnes Pockels (1862–1935)

Katharina Al-Shamery

Agnes Pockels developed an instrument to study the interfaces of liquids, today known as a Langmuir–Pockels trough (or more often a Langmuir trough), and thus pioneered research into surface tension. She did not have any formal scientific training, neither a university degree nor even a baccalaureate, but has been honored as the first and (until today) only woman with an honorary PhD from the technical university Carolo-Wilhelmina of Brunswick, Germany, and the Laura-R.-Leonard prize of the Colloidal Society on the occasion of her 70th birthday.

"My lord, Will you kindly excuse my venturing to trouble you with a German letter on a scientific subject? Having heard of the fruitful researches carried on by you last year on the hitherto little understood properties of water surfaces, I thought it might interest you to know of my own observations on the subject. For various reason I am not in a position to publish them in scientific periodicals, and I therefore adopt this means of communicating to you the most important of them." This letter on her research, inititally triggered by observations on fatty dishwater 10 years earlier, was written by Agnes Pockels to Lord Rayleigh when she was 29 years old. Her letter was published as her first paper in Nature and was a turning point in her life.

Agnes Pockels was born in Venice on 14th February, 1862 as the elder of two children of German parents Alwine Pockels, born Becker, and Theodor Pockels, an officer who served in the Austrian army in northern Italy. Her younger brother Friedrich was born in Vincenza three years later. Malaria was widespread in this area. The whole family had severe health problems so that Pockels' father had to retire early and returned to Brunswick near the Harz mountains in 1865.

Agnes Pockels attended the local girls school with a major focus on German, religion and languages. When she finished school women were still not admitted at the university. Later, her parents did not allow her to enter university as she had to take care of her ailing father. The household was her major occupation. Her brother took a different career. After finishing school he studied physics, first at the technical university Carolo-Wilhelmina of Brunswick, the Albert Ludwigs university of Freiburg and then at the Georg August university of Göttingen. He first received an appointment as a professor in Dresden and later in Heidelberg. His work focussed

European Women in Chemistry. Edited by Jan Apotheker and Livia Simon Sarkadi
Copyright © 2011 WILEY-VCH Verlag GmbH & Co. KGaA, Weinheim
ISBN 978-3-527-32956-4

Agnes Pockels

on the influence of electrostatic fields on optical properties. The "Pockels-effect" is text book knowledge. The Pockels cell is an important component of modern laser systems.

With her brother Friedrich Agnes could discuss physics, her favorite subject. In 1880, at the age of 18, she observed, while working in the kitchen, that the surface tension of water changes with the dissolution of impurities from solid bodies dipped into it. Plinius the elder and Plutarch, as well as Benjamin Franklin, had already described the interaction of oil with water sufaces. However, there was no experimental method known to study the phenomena in detail. In 1882 Agnes Pockels developed a trough in which one could change the surface with a slider and measure the surface tension quickly and precisely with an accurate balance. She permanently refined the apparatus. Later her niece, Anna Pockels, Friedrich Pockel's daughter, described it as made from a tin from "Liebig's Fleischextrakt", the pharmacy balance of her grandfather which had a a ring of platinum wire insted of a scale pan. The ring dipped into a channel containing the liquid to be investigated. Irvine Langmuir (Nobel laureate for Chemistry in 1932) later refined her set-up and developed, together with Katherine Burr Blodgett, a method to produce monolayers on solids and liquids (the Langmuir–Blodgett technique). Soon after she started with her first experiments Agnes Pockels' brother began his physics studies at the university in 1883 and provided her with text books and publications so that she could teach herself the necessary knowledge of physics. However, the professors in Göttingen were not interested in her work. Without a direct contact with scientists it was not possible for her to publish her results.

The breakthrough for Agnes Pockels came in 1891 when she wrote a 12 page letter to Lord Rayleigh (1842–1919) summarizing her work of nearly 10 years. Lord Rayleigh had just published a paper in the *Proceedings of the Royal Society* about his observations on the film formation of olive oil on water. Agnes Pockels read about his work in a summary of this paper in the journal *Naturwissenschaftliche Rundschauen*. Lord Rayleigh was not interested in taking credit for other people's work and thus submitted her letter to *Nature*, where her work was published under the title "Surface Tension" with Rayleigh's introductory remarks:

I shall be obliged if you can find space for the accompanying translation of an interesting letter which I have received from a German lady, who with very homely appliances has arrived at valuable results respecting the behaviour of contaminated water surfaces. The earlier part of Miss Pockel's letter cover nearly the same ground as some of my own recent work, and in the main harmonize with it. The later sections seem to me very suggestive, raising, if they do not fully answer, many important questions. I hope soon to find opportunity for repeating some of Miss Pockels' experiments. – RAYLEIGH, March 2, 1891.

Further papers followed in *Nature* between 1892 and 1894. Now the German physicists started to acknowledge her work. Agnes Pockels travelled frequently the 100 km to Göttingen. However, she could not make use of the offer to work in the physics laboratories as her parents were permanently ill and had to be taken care of. Instead she had to find time at home to carry out experiments, and she published further papers in the years 1898 to 1902 on adhesion of liquids to glass, contact angles of saturated liquids at crystals and the surface tension of emulsions and solvates. In 1900 her brother moved to Heidelberg, more than 400 km from Brunswick. After 1902 Agnes Pockels turned more toward theoretical work. Her broad scientific interests are documented in her translations of Georg Howard Darwin's book on "The Tides and Kindred Phenomena in the Solar System" or in her philosophical paper on "The Abritrariness of the World". In 1906 her father died, followed in 1914 by her mother. But the worst loss for her was the death of her brother in 1913. For this reason, and also because of World War I, her own health problems and her worsening eyesight she was less and less able to obtain access to the current publications. Nevertheless, she still published five papers by 1918 and two further papers thereafter. Her living costs were financed by American relatives, otherwise her work would not have been possible. In 1932, the year of her 70th birthday she received an honorary PhD in engineering from the faculty of mathematics and physics of the technical university Carolo-Wilhelmina of Brunswick and the Laura R. Leonard prize of the Colloidal Society on the occasion of a conference of the Physical Society in Brunswick. The apprasial for this occasion was written by Wolfgang Ostwald in the *Kolloid-Zeitschrift* (a journal on colloids) who recognized her as the founder of quantitative film research.

Agnes Pockels died in 1935. Though she was well known during her lifetime Agnes Pockels has been nearly forgotten today while her brother is still famous.

Literature

Beisswanger, G. (1991) Agnes Pockels (1862–1935) und die Oberflächenchemie, *Chemie in Unserer Zeit*, **2**, 97.

Ostwald, W. (1932) The work of Agnes Pockels about interfaces and films, *Kolloid Z.*, **58**, 1.

Pockels, A. (1981) Surface Tension, *Nature*, March 12.

Pockels, A. *Diaries*, Archive TU Brunswick

Poggendorff, J.C. (1938) *Biographisch-Literarisches Handwörterbuch* (*Biographical Literary Dictionary*), vol. VI: 1923–1931, Berlin.

Marie Skłodowska-Curie (1867–1934)

Renate Strohmeier

■ Marie Skłodowska-Curie, by far the most famous woman in science, not only the first woman to win a Nobel Prize but also the only woman to win it twice.

Although Marie was already a member of the Swedish, Czech and Polish Academy of Sciences and several other prestigious societies, her nomination for the membership of the French Académie des Sciences was declined in January, 1911. It took a surprisingly long time until the first woman, Marguerite Perey, discoverer of Francium, was admitted (1962). Marie's nomination provoked a smear campaign in the French press. Prejudice against academic women as well as xenophobia motivated absurd accusations, by which she was deeply hurt. In the same year a second campaign was started. This time concerning her love affair with Paul Langevin, family friend and Pierre's former student. Although Paul lived separated from his wife, Marie was accused of destroying his family and the affair went as far as threatening Marie's life. Actually two duels took place, albeit without deadly outcome. The affair even affected the granting of Marie's second Nobel prize. A member of the Swedish Academy of Science, Svante Arrhenius, asked her not to come to Stockholm to accept the prize. However, Marie was not dispirited and went to Stockholm with her sister Bronia and her daughter Irène to receive the Nobel Prize in Chemistry in December 1911. This time she gave the Nobel lecture herself and made clear, which part in the collaboration with Pierre was hers.

Jointly with Henri Bequerel and Pierre Curie, her husband, she was awarded the 1903 Nobel Prize for physics "in recognition of the extraordinary services they have rendered by their joint researches on the radiation phenomena discovered by Professor Henri Becquerel". In 1911 she was the sole Nobel Prize winner in Chemistry "in recognition of her service to the advancement of chemistry by the discovery of the elements radium and polonium, by the isolation of radium and the study of the nature and compounds of this remarkable element". (Citations of the Nobel Committee).

When Mary Skłodowska came to Paris in 1891, many years of deprivation were behind her and more years of hard work under poor conditions lay before her. She changed her name to the French "Marie" and began her studies at the Sorbonne's

European Women in Chemistry. Edited by Jan Apotheker and Livia Simon Sarkadi
Copyright © 2011 WILEY-VCH Verlag GmbH & Co. KGaA, Weinheim
ISBN 978-3-527-32956-4

Marie Curie official Nobel prize photograph, 1911.

faculty of science as one of 23 women among 1825 students. Some of her teachers were France's leading scientists, such as the mathematician Paul Painleve and the physicist Gabriel Lippmann. Two years later she passed her physics exam as the best student of her class. In the following year she was second in a degree in mathematics. When Marie Skłodowska met the 35 years old internationally known physicist Pierre Curie in 1894, each found in the other a complementary mind and personality, determined to dedicate their entire lives to science. The fateful encounter changed Marie's plans to go back to her country to work for an independent future for Poland.

In the then Russian-dominated Poland, Mary and her sisters first went to a school were Polish language and culture were taught clandestinely, while the official syllabus was dictated by the Russian occupants. Like their parents, the children of the Skłodowska family were intensely patriotic. Mary was a believing catholic, but when her mother and a sister died in 1878 she lost her faith and became agnostic. Mary, as well as her sister Bronia, was determined to study at a university, but at that time advanced study was not possible for women in Poland. As their father was not in a position to support an education abroad, they made an agreement: Bronia would go to Paris first and study to become a medical doctor and Mary would work as a governess in Poland and support her. As soon as Bronia was earning money she would put Mary through University. The plan worked well and Mary went to Paris.

When Marie met Pierre he was head of a laboratory at the School of Industrial Physics and Chemistry, where engineers were trained. At the age of 21, he and his brother Jacques made the very important discovery of piezoelectricity. The two brothers also invented the piezoelectric balance, a tool, which played a major role

Marie and Pierre Curie in their laboratory.

in later investigations of radioactive elements. When Jacques left in 1883 to become the head lecturer in mineralogy at the University of Montpellier, Pierre pursued the investigation of crystals and magnetic properties of bodies in relation to temperature. This led to his doctoral thesis in 1895, which contained the evaluation of the connection between temperature and magnetism that is now known as Curie's Law. He was described as "a serious idealist and dreamer whose greatest wish was to be able to devote his life to scientific work". Marie became his marital and scientific partner in a collaboration on equal terms, sharing both work and credit. H. M. Pycior, who analyzed their collaboration wrote: "In the case of Marie and Pierre, the contrast is (also) that of a thinker-dreamer (Pierre) who reveled in broad reflection on nature and that of a thinker-doer (Marie) whose *steadfast need for clarity* helped to bring such reflection to fruition".

They were married in 1895, the Curies agreed that Marie would never give up scientific work and teaching. With the support of Eugene Curie, Pierre's father, who moved in with the couple after his wife's death, and by hiring Polish nurses and later governesses, they established a lifestyle in which there was time for science and family.

Over the first period after their wedding Marie passed her teacher's diploma and completed investigations on the magnetic properties of various steels on behalf of the Society for the Encouragement of National Industries. As Marie was determined to continue doing research she decided to study for a doctorate. Fascinated by the recent discovery of X-rays by Wilhelm Conrad Röntgen and Bequerel's observations on 'uranium rays', both she and Pierre considered them to be a good subject for her thesis. Marie's ambition was "to determine the intensity of the radiation (of various substances), by measuring the conductivity of the air exposed to the action of rays". She was fortunate to have the equipment to measure weak electric currents with great precision simply because the electrometer, invented by Pierre and his brother and unused for years, was available in the laboratory. She later explained: "One of the most important properties of the radioactive elements is that of ionizing the air in their vicinity". Marie had to perform her experiments under difficult conditions and very poor laboratory arrangements. Despite her teaching obligations and a three month old daughter, within a short period of time she made some absolutely revolutionary discoveries:

- the intensity of the radiation is proportional to the quantity of the active element in the probed sample,
- the radiation is not effected by external factors such as light or temperature, which led to the conclusion that
- the emission of rays is a property of the atom itself, independent of the chemical or physical state.

As she was not a member of the *Académie des Sciences,* her results were presented on April 12, 1889, by her former teacher Gabriel Lippmann. A paper titled "Rays Emitted by Compounds of Uranium and of Thorium" was published 10 days later. The Curies refused an offer to register a patent for their findings. They were convinced that scientific results belong to humanity.

Her testing of pitchblende revealed that its radioactivity was four times greater than justified by its uranium content. By discussing the discrepancy, which was similarly seen in chalcolite, Marie and Pierre hypothesized that the radiation came from a new chemical element. At that time (almost three years after their wedding) Pierre abandoned his own research into crystals and symmetry in nature and the Curies became partners in the study of radioactivity, determined to find the new chemical element.

Trying to separate the different elements of pitchblende by chemical methods they found that the strong activity came with the fractions containing bismuth or barium. When Marie continued the purification of bismuth, a residue with greater activity was left. In June 1898 they had isolated a substance akin to bismuth, which was 330 times more active then uranium. Although spectrographic proof failed, Henri Becquerel presented her findings "On a New Radioactive Substance Contained in Pitchblende" to the Académie des Sciences in July 1898. Here, for the first time, the term *radioactivity* was used for the spontaneous emission of radiation. The Curies suggested that the new element should be called *polonium* after the name of Marie's native land. In July of the same year the *Prix Gegner* of the Académie des Sciences was awarded to Marie Curie for her work on magnetic properties of steel and on radioactivity. Less than 6 months later, Marie and Pierre reported on an additional radioactive element found in pitchblende which they named *radium,* the most powerful radioactive substance ever discovered. This time a new spectral line for the element could be demonstrated by E.A. Demarçay. As proof of the existence of polonium and radium, the Curies had to isolate them in demonstrable amounts and to determine their atomic weight.

Early in 1899 Pierre began to place emphasis on the physical effects of radioactivity in cooperation with Georges Sagnac and André-Louis Debierne, while Marie devoted herself totally to the chemical isolation of radium. To accomplish the task they would need enormous amounts of costly pitchblende. With the assistance of the Vienna Academy of Sciences, Marie got several tons of slag from the Joachimsthal mine in Bohemia, which was even more active than the original pitchblende. As the present laboratory was too small for their purpose, Pierre's school provided a large draughty shed for the very exhausting tedious work of separation and analysis. She writes: "Sometimes I had to spend a whole day stirring a boiling mass with a heavy iron rod nearly as big as myself. I would be broken with fatigue at day's end". The German chemist Wilhelm Ostwald visited the Curies to see how they worked. He later wrote: "At my earnest request, I was shown the laboratory where radium had been discovered shortly before... It was a cross between a stable and a potato shed, and if I had not seen the worktable and items of chemical apparatus, I would have thought that I had been played a practical joke". With the financial support of the French Academy of Science and only one technician, after four years of hard work, she had produced a suitable radium sample of one decigram to calculate the atomic weight. Marie detailed the findings in her doctoral thesis *Recherches sur les Substances Radioactives*, which she presented on June 25, 1903. Two members of the examination committee, Gabriel Lippmann and Henri Moissan, were future Nobel Prize winners. Marie's dissertation was translated into five languages and reprinted 17 times in scientific journals.

Early in 1903, their first health problems appeared. Marie and Pierre ignored all obvious signs of impaired health due to contact with radiation. Becquerel, as well as Pierre Curie and other scientists working with radioactive substances, reported burn-like damage of the skin and Pierre's latent health problems contributed to extremely painful rheumatism. In retrospect, the careless handling, like demonstrating how a radium salt in solution illuminates the darkness of a garden party, strikes us as incredible. To date the dangers and long-term effects of radioactivity are not always taken very seriously.

In mid-November 1903 the Curies got the message from Stockholm that they had won half of the Nobel Prize in Physics. Henri Becquerel was awarded the other half for his discovery of spontaneous radioactivity. They were not able to go to Sweden to receive the prize. Both had serious health problems and it was two years before they went to Stockholm, in June 1905, when Pierre gave the Nobel lecture.

In December 1904, Ève, her second daughter, was born. She later wrote the rather romantic biography of *Madame Curie,* her mother. She was not even two years old, when her father died in a traffic accident on April 19, 1906. Marie Curie not only lost her beloved husband but also her scientific partner. She succeeded Pierre in his position at the Sorbonne, first as a lecturer, and then, two years later, as professor. She was the first women ever appointed to teach at the University of Paris. Invited by Ernest Rutherford, Marie Curie agreed to establish a unit for the activity of a quantity of a radioactive substance, which became necessary because radium was increasingly used in medicine, industry and research. On the occasion of the World Congress on Radiology and Electricity in Brussels a commission of ten scientists, including Marie, decided to name the measurement "Curie". In 1975 the "Becquerel" replaced the Curie as the official radiation unit.

Returning from Sweden her state of health declined. She became more and more depressed and suffered from pyelonephritis. It took her almost two years to recover and return to work. In 1913 she went to Warsaw to inaugurate the new radium institute, which was built in her honor.

When World War I began, Marie immediately started organizing the equipment of vans with X-ray apparatus to be used as mobile field devices for locating metal splinters in wounded soldiers. With the support of the French women's league she installed the first mobile radiation van and trained young women in X-ray technology. With private donations and the help of *Le Patronage National des Blessés* about 200 radiology cars were installed. Throughout the war she and her daughter Irène trained technicians and also served themselves at the front.

After the war her Radium Institute opened. When Marie Curie revealed to the American journalist Marie Melony the moderate means available to equip it, Miss Melony started a fund-raising campaign to enable her to buy one gram of radium at the price of 100 000 US$. The chancellor of the Sorbonne, Paul Appell, had to persuade Marie to go to America to accept the money, because she still avoided publicity. When she returned to Paris, a great gala at the Opera was performed in her honor and Marie Curie was celebrated as a modern Jeanne d'Arc, after being dragged through the mud ten years before, as stated in the Nobel Institute essay about Marie and Pierre Curie.

When, in 1922, the League of Nations established a new commission for intellectual cooperation, Marie Curie was elected as one of 12 members. During the 12 years of her active membership she served for a time as vice-president. She became involved with the foundation of an international scientific bibliography and establishment of guidelines for the award of international research grants amongst others. As director of the Radium Institute she encouraged particularly women and students from foreign countries. In 1931, 12 out of 37 scientists were women, amongst them Ellen Gleditsch from Norway, Eva Ramstedt from Sweden, Marietta Blau from Austria and Marguerite Perey from France.

Marie Curie died on July 17, 1934 in Sancellemoz, near Passy, France, from leukemia. She became a victim of her unconcerned exposure to radium and X-rays. She did not live to know that her daughter Irène and her son-in-law Frédéric Joliot followed her in winning the Nobel Prize in Chemistry in 1935, but she experienced their discovery of artificial radioactivity.

Timeline for Marie Curie:

1867 November 7th born in Warsaw, in the then Russian-dominated Poland. Parents: Wladislaw Skłodowska, teacher of mathematics, and Bronislawa (née Boguska), principal of a preparatory school for girls.
1891 Marie went to Paris and began her studies at the Sorbonne
1893 Physics exam as the best student of her class
1894 Mathematics exam as the second-best student
1894 met her future husband Pierre Curie
1895 married Pierre Curie
1897 Irène and 1904 her second daughter Eve were born
1898 the *Prix Gegner* of the Académie des Sciences was awarded to Marie
1903 doctoral thesis *Recherches sur les substances radioactives*
1903 first Nobel prize, shared with Henri Becquerel and Pierre Curie
1906 Pierre Curie died
1908 Marie became Professor at the Sorbonne

1911   second Nobel prize for Marie Curie

1934, July 17th Marie died in Sancellemoz, near Passy, France

Literature

Brian, D. (2005) *The Curies. A Biography of the Most Controversial Family in Science*, John Wiley & Sons, Hoboken, NJ, USA.

Curie, È. (1952) *Madame Curie*, Fischer Verlag, Frankfurt am Main.

Fröman, N. (1996) *Marie and Pierre Curie and the Discovery of Polonium and Radium*, Lecture at the Royal Academy of Sciences in Stockholm, Sweden, on February 28, 1996. (http://www.nobel.se/essays/curie/index.html)

Pycior, Helena M. (1996) Pierre Curie and "his eminent collaborator Mme Curie". Complementary partners, in *Creative Couples in the Sciences* (eds. H. M.. Pycior, N. G. Slack, and P.G. Abir-am) Rutgers University Press, New Brunswick, N. J.

Quinn, S. (1999) *Marie Curie. Eine Biographie*, Insel-Verlag, Frankfurt am Main.

Clara Immerwahr (1870–1915)

Marianne Offereins

■ Clara Immerwahr studied chemistry and attained her Doctor's degree on the solubility of various metal salts. She married Fritz Haber who later became a Nobel laureate.

Because she could not live with the idea that her husband developed poisonous gases for chemical warfare, she took her life in 1915 with a shot from her husband's duty pistol.

On June 21, 1870, Clara Immerwahr was born at the Polkendorf estate in Breslau in Silesia (now Wrocław, Poland). Clara was the youngest of four children of Dr. Philip Immerwahr and Anna Krohn Immerwahr (besides Clara, there were two girls, Elli and Rose, and a boy, Paul). The children grew up in a wealthy highly-cultured, open and liberal family.

Clara Immerwahr.

European Women in Chemistry. Edited by Jan Apotheker and Livia Simon Sarkadi
Copyright © 2011 WILEY-VCH Verlag GmbH & Co. KGaA, Weinheim
ISBN 978-3-527-32956-4

Clara received her early education at home, together with her brother and sisters. She was a diligent student and soon there was competition between Clara and her older brother Paul. From the winter semester of 1877 the three sisters Immerwahr visited the *Höhere Töchterschule* of Fräulein Krug in Breslau. In the summer the family lived on the ancestral estate, where the children were taught by various governesses. From the beginning of her school time Clara showed a great interest in the sciences and one could annoy her deeply by pointing to her '*künftige weibliche Wirkungskreis*' (future feminine occupations). During dancing lessons in Breslau Clara met Fritz Haber. The two felt attracted to each other, but Clara's best friend called Fritz "*erzgescheit, aber gespreitzt und eitel*" (extremely clever, but pompous and vain). Fritz wanted to get married as soon as possible, but the mutual parents thought that first he had to be able to earn a living, and Clara was not convinced of her feelings for Fritz. Fairly soon thereafter her brother Paul attained his doctor's degree. From that moment Clara wanted to be socially independent by having a university education. To achieve this, she followed the route many German women and girls of her age were taking: through the *Lehrerinnenseminar* (teacher's seminary). Clara entered the *Seminar* in Breslau. The headmistress of the *Seminar* soon noticed Clara's scientific interest and gave her *Unterhaltungen über die Chemie* by Jane Marcet (*Conversations on Chemistry* (1806), still popular more than 70 years after its release). This book definitely confirmed Clara's love of chemistry. Clara finished the final exams of the teachers training. Now she had to face obstacles and prejudices. Together with her father she decided to take private lessons first, in which *Geheimrat* Albert Ladenburg supported and encouraged her. Finally – in 1896 – Clara, and a few other girls were accepted as auditors at the lectures at the university. Of course, chemistry was her great interest, and, especially, the chemical experiments gave her great pleasure, despite the ridicule, the opposition and the fact that she and her fellow girl students were ignored by male professors and fellow students.

At the end of the winter semester of the academic year 1896/97 department head Küster left the University of Breslau. He was succeeded by Richard Abegg, who was a college friend of Haber's, Clara and Abegg got along very well. In the same year came an announcement of the *Kulturministerium* that would be of great importance for the progress of Clara's education, being an auditor meant that she had the same rights as a regular student. Immediately Clara and Abegg went looking for a suitable PhD topic. They chose to investigate the solubility of various heavy metal salts.

For her preliminary research Clara traveled to the *Bergakademie* in Clausthal, where she was supervised by Professor Küster. She often had doubts about her own capabilities and her work.

By her measurements and her experience, she discovered that the potential measurements of Gauss were imperfect, or rather, random values. For the first time, an investigation of Clara was published. The *Zeitschrift für Anorganische Chemie* published her article '*Potentiale von Cu-Elektroden in Lösungen wichtiger Cu-Niederschläge*'. On June 28, 1900, Clara made a request for admission to the doctorate. Her request was supported by 31 teachers. Clara was accepted. Her research was in the field of physical chemistry, her thesis was titled: "*Löslichkeitsbestim-mungen schwerlöslicher Salze des Quecksilbers, Kupfers, Bleis, Cadmiums und Zinks*".

She dedicated her dissertation to "*dem lieben Vater*". The verdict on her oral exam was unanimous: *magna cum laude*. On December 22, 1900 the oral defense took place. Immediately after that Clara Immerwahr was, as the first woman in Breslau, formally granted the doctorate. Thus Clara Immerwahr became the first woman in Germany to attain the Doctor's degree in chemistry. After her promotion Clara became assistant to Professor Abegg, at that time the highest attainable academic position for a woman.

In 1901, at the congress of the Deutsche *Gesellschaft für Electrochemie*, in honor of the deceased Robert W. Bunsen, Clara met Fritz Haber again. For the second time in her life he proposed to Clara and after an initial hesitation, this time she accepted. In August Clara Immerwahr and Fritz Haber were married. The couple settled in Karlsruhe, where Fritz had good connections with the chemical industry.

Soon after the marriage, Clara became pregnant, and, after a difficult pregnancy, son Hermann was born. After the delivery Fritz was in bed as well, with stomach complaints.

Shortly after the birth of Hermann the marriage began to show its first cracks and Clara furnished a room of her own. Yet Fritz dedicated his book '*Thermody-namik technischer Gasreaktionen*', with which he was widely recognized, to: "*Meiner lieben Frau Clara Haber Dr. Phil. zum Dank für stille Mitarbeit zugeeignet*" (Dedicated to my dear wife Dr. Phil. Clara Haber, to thank her for her quiet cooperation).

After the discovery of nitrogen as a fertilizer, Fritz threw himself into research on the synthesis of ammonia. He wanted to make a career: "*Nicht möglichst wenig ausgeben, sondern möglichst viel verdienen*" (Not spend as little as possible, but earn as much as possible). In the beginning Clara contributed much to the work of her husband, however, without having her name mentioned as a collaborator.

From the moment the war broke out on August 1st, 1914, the research on ammonia as a base for fertilizer moved more and more to the manufacture of explosives and of poison gases, such as chlorine, phosgene, and so on, for the war industry.

Meanwhile, both Clara and Fritz went more and more their own ways. Fritz was a workaholic who did not want to give time to his wife and his child, Clara held lectures for women on 'Chemistry and Physics in housekeeping'. By this she expressed her anti-militarist position, against the patriotic attitude of her husband and his institute. Clara resisted fiercely her husband's research on chemical warfare. She described the research as "*eine Perversion der Wissenschaft*" (a perversion of science). She could not convince him of the disastrousness of his research. When the catastrophic effects of chlorine as a war gas, which was developed by Haber, were clearly shown – in one of the first attacks the French alone lost 18 000 people – she could no longer live with that responsibility. On the morning of May 2, 1915, she took her life with a shot through the heart with her husband's duty pistol.

> *(...) Es war stets meine Auffassung vom Leben, daß es nur dann*
> *wert gewesen sei, gelebt worden zu sein, wenn man alle seine*
> *Fähigkeiten zur Höhe entwickelt und möglichst alles durchlebt*
> *habe, was ein Menschenleben an Erlebnisse bieten kann. (...)*
> (...) It was always my understanding of life, that it is only
> worth living, if you've developed all your skills to the highest
> possible level, and possibly have experienced everything that
> a lifetime can provide. (...)
> Clara Haber-Immerwahr in retrospect. 1909.

Since November 15, 2000, at the University of Dortmund, there has been the Clara-Immerwahr-mentoring project, intended to encourage women to study chemistry.

Literature

Leitner, G. von (1994) *Der Fall Clara Immerwahr. Leben für eine humane Wissenschaft*, München.

Molenaar, L., and Kooiman, P. (1986) *Chemie en Samenleving. Van Kleurstof tot Kunstmest*, Maastricht/Brussels.

Offereins, M. I.C. (1996) *Vrouwenminiaturen. Biografische schetsen uit de exacte vakken*, Utrecht.

Ogilvie, M. and Harvey, J. (eds) (2000) *The Biographical Dictionary of Women in Science. Pioneering Lives from Ancient Times to the Mid-20th Century*, Routledge, Cambridge, MA/London.

Strohmeier, R. (1998) *Lexicon der Naturforscherinnen und Naturkundigen Frauen Europas. Von der Antike bis zum 20. Jahrhundert*. Thun und Frankfurt am Main.

Maria Bakunin (1873–1960)

Marco Ciardi and Miriam Focaccia

■ Maria Bakunin – to her friends "Marussia", the name under which she even published some of her works – made a significant contribution to progress in the field
of chemistry, and also to female emancipation in the difficult times when women
in Italy were beginning, gradually and laboriously, to make a place for themselves
in the ranks of the "big" sciences at university: chemistry, mathematics and
physics. In 1912 she was assigned a lecturing post in chemistry at the *Scuola Politecnica*, marking a departure from the nineteenth-century tradition of women only teaching the 'natural sciences'. With a strong and decisive character, Maria was
a central figure in Neapolitan circles at the time, both for her ground-breaking research in stereochemistry and photochemistry and for her chairmanship of various scientific institutions and participation in the intellectual life of the city. In 1947
she was the first woman to be elected a member of the *Accademia Nazionale dei Lincei*, in the physical sciences class.

Maria Bakunin, the third child of the Russian philosopher and revolutionary
Michail Bakunin, was born in Krasnojarsk in Siberia on 2 February, 1873. When
her father died in Berne in 1876, she went with her family to Naples, where they
had maintained the many bonds her father had created with the city. After completing her studies at the classical *lycée* (where women had only recently been admitted), with her brother Carlo and her sister Giulia Sofia she enrolled on the
chemistry degree course at the university. In 1895 she graduated with a thesis entitled *On phenyl-nitrocinnamic acids and their stereometric isomers*. Her sister Giulia
Sofia graduated from the same university in medicine and surgery in 1893.

Although, on the one hand, from 1890 onwards she numbered among the pre-
parers at the Neapolitan Institute of Chemistry directed by Agostino Oglialoro-To-
daro (later her husband), on the other hand she was perfectly at her ease in Rome
in 1896 among the Italian chemists who had gathered to celebrate Stanislao Cannizzaro's 70th birthday.

In fact it was Cannizzaro, along with Emanuele Paternò, whose good opinion of
her led to her studies being recognized in the form of the coveted Academy prize
for physics and mathematics in Naples in 1900.

European Women in Chemistry. Edited by Jan Apotheker and Livia Simon Sarkadi
Copyright © 2011 WILEY-VCH Verlag GmbH & Co. KGaA, Weinheim
ISBN 978-3-527-32956-4

Maria Bakunin (Libera Universitá delle donne).

In 1902 she was one of the attendees at the 1st National Conference on Applied Chemistry, a conference convened by the Industrial Chemistry Association in order to try to establish an Italian chemical company. This intention was also expressed through the publication of the periodical *Chemistry and Industry*.

In 1909 she began teaching applied chemistry at the *Scuola Superiore Politecnica* in Naples and in 1912 she won the open competition for the Chair in Applied Technological Chemistry at the same establishment. In 1921 she became the president of the Neapolitan branch of the Italian Chemistry Association. In 1928, as well as being an authoritative member of the governing body of the Chemistry Commission of the CNR (*Consiglio Nazionale delle Ricerche* – National Research Council), she was appointed one of the 13 members of the so-called 'Aromatic Hydrocarbons Commission' (whose aim was to procure benzene, toluene and other additives) by the President, Nicola Parravano. This commission was then broken up in 1930 by Parravano in order to make room for the members of the Combustion Fuels Commission.

In 1940, she took the Chair of Organic Chemistry at the science faculty of the University of Naples, where she worked until 1947. She was part of the Chemistry Committee of the NCR during the period of reconstruction in 1945 and 1946.

Maria Bakunin carried out wide-ranging and in-depth research into indones, research that can be linked to her first studies of the geometric isomerism of nitrocinnamic and oxy-cinnamic acids, on which she focused her energies right from her undergraduate thesis. Alongside these ground-breaking studies we can also note her research into the make-up of picrotoxin, the esterification of phenols, the catalyzing effect of certain colloidal solutions in organic syntheses, and also contributions to the field of applied chemistry that led to the preparation of some important medicinal products.

As far as organic synthesis is concerned, Bakunin will be remembered for having introduced an original method of preparing indones, anhydrides and ethers based on the use of phosphorous pentoxide in chloroform (1900). The studies she carried out in collaboration with Peccerillo and published in the *Gazzetta* (1933–1935) are often cited.

Also interested in earth sciences, in 1906 Bakunin took part in an observation group studying the eruption of Vesuvius, while from 1909 to 1910 she took on a project for the Italian education ministry to compile a geological map of Italy. For this project, she paid particular attention to the oil shale and ichthyolithic deposits typical of the South Tyrolese Dolomites and the Picentini mountains in the Salerno area.

From 1911 to around 1930, she worked as a consultant for the Giffuni district council and for companies from the same area hoping to exploit the local ichthyol mines. This activity helps us to see how active Bakunin was in working towards the dream of driving industrial development in the areas around Naples, thus making it a stronger presence in the crucial economic processes of the country as a whole. In this light, the close collaboration with her pupil Francesco Giordani, the future famous technocrat with whom she maintained a close working relationship in years to come, appears particularly significant.

In Maria, vast knowledge, organizational capacities and teaching abilities were never separate from more human qualities like courage and moral rigidity. In 1938 Maria intervened to prevent her nephew, the mathematician Renato Cacciopoli, from being arrested for anti-fascist activities. Maria was not at all frightened of the Fascist authorities or the adversities of war: when her house was burned down by the Germans, she moved to a big, empty hall at the university, which soon became a veritable Noah's ark where she continued to provide hospitality to all those who turned to her. For the whole duration of the war, she never left the Institute of Chemistry; indeed she saved it from the Allied troops that tried to occupy it for military use.

Maria was very close to the philosopher Benedetto Croce and together they rebuilt the *Accademia Pontaniana* after the second world war. In 1944, after Fascism was defeated, she was even elected president of the *Accademia*, of which she had been a member since 1905. One of Bakunin's greatest achievements in this role was the restoration of the precious library, which had been destroyed by fire. In the post-war period Maria also helped to rebuild the university, working alongside the chancellor Adolfo Omodeo.

She published her research in the *Gazzetta Chimica Italiana*, the *Annali di Chimica Applicata* and in the *Proceedings of the Società di Scienze, Lettere ed Arti* in Naples and *the Academy of Science* in Bologna.

She died in 1960 at her home of Mezzocannone, in Spaccanapoli – just a stone's throw from the chemistry institute – at the age of 87.

In the Commemoration read by Rodolfo A. Nicolaus, a pupil of Bakunin's at the University of Naples, at the *Accademia delle Scienze di Napoli* in 1961, Maria was described as a professor who was "authoritarian but with great charm and prestige" and also a real character full of personality. In fact, for many years she added sparkle to the world of chemistry in Naples; with elegant femininity she was the perfect host and her convivial receptions – her *salon* – were always open to the important figures in the local and national spheres of culture and knowledge.

Literature

D'Auria, M. (2009) La nascita della fotochimica in Italia. Il ruolo di Maria Bakunin. in *Atti del XIII Convegno di Fondamenti e Storia della Chimica* (Roma, 23–26 Settembre 2009), (ed. Calascibetta, F.) Accademia Nazionale delle Scienze, Roma, pp. 161–172.

Fascicolo Personale di Maria Bakunin, Archivio Centrale dello Stato, Ministero Pubblica Istruzione, Direzione Generale Istruzione Universitaria, Fascicolo professori Universitari, III serie (1940–1970), Da Bay a Bak, B. 28.

Malquori, G. (1961/1962) Marussia Bakunin. *Atti dell'Accademia Pontaniana*, **11**, 393–399.

Maria Bakunin. Commemorazione letta dal socio Rodolfo A. Nicolaus. (1961) *Rendiconto dell'Accademia delle Scienze Fisiche e Matematiche della Società nazionale di Scienze, Lettere ed Arti in Napoli*, ser. IV, **28**,15–21.

Mongillo, P. (2008) *Marussia Bakunin. Una donna nella storia della chimica*, Rubettino, Napoli.

Nicolaus, R.A. (1988) Bakunin Marussia, in *Dizionario Biografico Degli Italiani*, vol. 34, Istituto della Enciclopedia Italiana, Roma, pp. 223–224.

Nicolaus R.A. (1960) Maria Bakunin. *La Chimica e L'Industria*, **42** (6), 677–678.

Patuelli, F. (2008) Bakunin Maria. in *Scienza a Due Voci. Le Donne Nella Scienza Italiana dal Settecento a Novecento*, (eds V. Babini and R. Simili) (http://scienzaa2voci.unibo.it)

Simili, R. (2008) In punta di penna, in *La Scienza nel Mezzogiorno Dopo l'Unità d'Italia*, vol. 1, Rubettino, Napoli, pp. 27–89.

Margarethe von Wrangell, Fürstin Andronikow (1876–1932)

Marianne Offereins

Margarethe von Wrangell refused to surrender to the boredom which was – willingly or not – the fate of women in her position, and became the first female professor in Germany. Her work on the subject of soil fertilization, without the help of foreign phosphates, was vital to the impoverished Germany.

Margarethe, called Daisy by her family, descended from an ancient Baltic noble family and was born on Christmas Day 1876 in Moscow as the third child of Baron Karl Fabian von Wrangell and his wife Ida. (In a number of biographies and collections her birth is given as January 7, 1877). During her early years she was exceptionally healthy and always in good humor. Throughout her life one of her characteristic features would remain that she never complained, but immediately knew how to react when something was wrong. When she was about three years old she had scarlet fever with life-threatening complications which were not recognized. Her health remained fragile, so her parents were advised not to expose her to too much education. She did, however, join her older brother and sister, who initially were taught at home by their mother. When her mother found she was pregnant

Margaret von Wrangell (http://margarete-von-wrangell.de/index.php?nav=4).

European Women in Chemistry. Edited by Jan Apotheker and Livia Simon Sarkadi
Copyright © 2011 WILEY-VCH Verlag GmbH & Co. KGaA, Weinheim
ISBN 978-3-527-32956-4

with her fourth child, Nicolai, Daisy's brother, went to school like the other boys his age, but for the girls a Russian teacher was appointed.

Father von Wrangell was an army officer, therefore the family had to move frequently. In Reval, the capital of Estonia (Tallinn today), where he was stationed from 1888, Daisy attended high school. She wrote about herself: *"Meine schönsten Erinnerungen sind mit der Mädchenschule der Baronesse von der Howen verknüpft; unter ihrer warmherzigen und vornehmen Leitung lernten wir die Freude am Lernen...."* (My finest memories are connected with the girl's school of Baroness von der Howen; under her warmhearted and distinguished direction we learned the joy of studying).

With the title *'Ausgezeichnet'* (excellent) she finished the Howen Schule in Reval, attained her teachers certificate at the boys school Nikolai I and, as a 19 year old, wrote novels which were published in the 'Revaler Zeitung' under the name Daisy Wrangell. As usual for a girl of her position, she played tennis and chess, she read Homer in Greek and Virgil in Latin, and she attended the 'compulsory' dances and spas for the upper classes. That kept her occupied, however, she found the balls and the spas annoying and philosophy and theology meant nothing to her. Eventually she discovered mathematics, which she thought "fabelhaft interessant" (utterly interesting).

As a child Daisy already had a strong desire for the outdoors and a great interest in nature and at the age of 26 she made her decision: she wanted to study science, "sollte es ihr letztes Armband kosten" (even if it would cost her last bracelet). Because of the death of her brother who studied chemistry in Zurich, she had to keep her plans a secret: the family was convinced that his death was due to excessive studying. Under the pretext to go for another rest cure, she took summer courses at Greifswald University. Only in 1904 could she officially register in Tübingen at the Eberhard-Karls University as a regular student. When this became known, a number of aunts broke off all contact with Daisy in relation to 'diese wahnwitzige Emanzipationsidee' (this mad emancipation idea). Not in the least disconcerted, Margarethe wrote back: *"Die Chemie hat so etwas Klassisches; (...) Man hört aus der Formel den ungeduldigen, leicht empfänglichen Herzschlag des Sauerstoffs; (...) man hört den schweren, trägen Blutstrom des Stickstoffs."* (Chemistry is something so classical (...) One hears from the formula the impatient easily susceptible heartbeat of oxygen; (...) one hears the heavy slow bloodstream of nitrogen).

The young agricultural chemist soon discovered the discipline that would be her life's fulfilment: plant physiology, and already one year after the beginning of her studies Daisy wrote home: *"Ich will mich jetzt gründlich auf Chemie, besonders die Organische legen, dann tüchtig Pflanzenphysiologie arbeiten und zum Schluß sehen, ob ich nicht was Neues über Stoffumsatz und Afbau in der Pflanze herauskriege."* (At first I want to apply myself thoroughly to chemistry, then work soundly on plant physiology and, finally, see whether I cannot find something new about metabolism and the structure of plants).

In 1909 she received her doctor's degree 'summa cum laude' on: *'Isomerieescheinungen beim Formylglutaconsäureester und seine Bromderivaten'*. (isomeric phenomena with the formyl glutaconic acid ester and its bromine derivatives). After this began her – what is called in the biography – 'Wanderjahre'. On the advice of her professor, Wilhelm Gustav Wislicenus, she left for England to do research with the Nobel Prize winner William Ramsay. Her work, in relation to radioactive thorium, was of such quality that Ramsay was impressed, and he complimented her on it. This praise from Ramsay would later open the doors for her to Marie Curie. After her apprenticeship in London she went, in 1910, to Strasbourg for a while, after which she went to Paris in 1911. The cooperation with her kindred spirit Marie Curie definitely established her name in the scientific world.

The family in Reval did not understand anything about her scientific achievements and wanted nothing better than that Daisy should come home. They already saw her put away in some harem as a white slave. Only when, in 1912, she was requested to take charge of a large agricultural experimental station in Estonia, did she turn back to Reval.

In 1922 she was invited to join the *Kaiser Wilhelm Institut für Physikalische Chemie und Elektrochemie* in Berlin. In 1923 she became the first woman in Germany who was appointed full professor, at the *Landwirtschaftliche Hochschule* in Hohenheim. For ten years, until her death, she would lead the institution, and with unrelenting enthusiasm she would lead many *'Diplomlandwirte'* to their license. She published on plant nutrition and soil enrichment (Phosphorsäureaufnahme und Bodenreaktion (1920), Über den Phosphorsäuregehalt natürlicher Bodenlösungen (1926), Über die Geschwindigkeit der Ionenaufname der Pflanzen (1928), and others). In 1926 she renewed contact with her childhood friend Vladimir Andronikow, who was also presumed dead, killed by the Bolsheviks. He wrote about this meeting: *"Ist es nicht wie ein Traum? Daisy, die kleine Kusine aus Reval, wird zum großen Professor in Deutschland! Die Elfe unserer Kinderphantasie gründet ein Pflanzenreich und regiert es mit einem Zauberstabe statt eines Szepters."* (Isn't it like a dream? Daisy, the little cousin from Reval, has become a great professor in Germany. The pixie of our children's fantasy grounds a country of plants and governs with a magic wand instead of a sceptre).

In 1928, at the age of 51, Daisy married Vladimir. The highest authorities gave her the license to continue working as a professor and as director of her institute after her marriage. (In that time women in Germany had to give up their work after their marriage, certainly when they worked in a civil service). The fact that she got permission, shows clearly how highly her scientific work was appreciated in government circles. A number of happy years followed, filled with work, traveling and social contacts. However, she could not enjoy her marriage for long. The scarlet fever from her childhood had left her with a kidney ailment, which now struck in full force. After a short illness she died at only 55 years old, on March 31, 1932. On April 4, 1932, she was buried.

Literature

Andronikow, Fürst Wladimir (1935) *Margarethe von Wrangell. Das Leben einer Frau 1876–1932; aus Tagebüchern, Briefen und Erinnerungen Dargestellt*, Albert Langen/ Georg Müller Verlag, München.

Angermayer, E. (1987) *Grosse Frauen der Weltgeschichte. Tausend Biographien in Wordt und Bild*. Neuer Kaiser Verlag – Buch und Welt, Klagenfurt.

Feyl, R. (1981) *Der Lautlose Aufbruch; Frauen in der Wissenschaft*. Verlag Neues Leben, Berlin.

Lina Solomonovna Shtern (also Stern, Schtern) (1878–1968)

Annette B. Vogt

■ Lina Solomonovna Shtern was a Jewish-Russian-Soviet biochemist. She was one of the founders of modern chemical physiology in the USSR. She did pioneering work on the hematoencephalic barrier, that is, the interface between the blood and the cerebrospinal fluid around the brain, which is called the blood–brain barrier. During her long life she published more than 500 scientific articles. She was the founder (in 1936) and chief editor (until her arrest) of the *Bulletin of Experimental Biology and Medicine*, and she was on the editorial boards of several other scientific journals.

Lina Solomonovna Shtern (also Stern, Schtern) was born on August 14th (Julian calendar) (August 26, Gregorian calendar) in 1878 in the town Libava (also Libau, Liepaja) in Latvia, Russia. Born in a Jewish family in tsarist Russia, educated and employed in Switzerland, later a professor in the Soviet Union, Lina S. Shtern was a cosmopolitan long before in the Soviet Union anti-semitic instigators denounced people like her as (bad) Jews using the word "cosmopolitan" instead of "Jew".

Shtern was born into the family of a successful merchant. One of her grandfathers was a Rabbi. In the family seven children grew up. She received a good education and attended the high school in Libava. Because of the discrimination of Jews in tsarist Russia, Jewish students had to study mostly in foreign countries, among others in Germany. Like many Jewish women from Russia, Lina Shtern went to Switzerland where she became one of the "Russian" women students at the University of Geneva. She studied medicine from 1898 until 1903, in 1903 she received her doctoral degree. Because of the hopeless job situation for women scientists and women Jews in Russia, Shtern stayed in Switzerland. After finishing her thesis she got an assistant position, in 1906 she received the "venia legendi" (Privatdozent) and finally, in 1917, she became Professor of Physiological Chemistry at the University of Geneva. She was a disciple of Jean-Louis Prevost Jr. (1838–1927) and worked together with his successor Federico Battelli (1867–1941). Until 1925 she made a remarkable scientific career as one of the first famous women scientists in Europe. In a short autobiographical sketch Shtern described herself as a feminist.

European Women in Chemistry. Edited by Jan Apotheker and Livia Simon Sarkadi
Copyright © 2011 WILEY-VCH Verlag GmbH & Co. KGaA, Weinheim
ISBN 978-3-527-32956-4

Lina Shtern (ca. 1930)

Attracted by the socialist experiment in the Soviet Union, Shtern decided in the mid-1920s to move to the Soviet Union. From 1925 onwards she lived and worked in Moscow. She became a full Professor of Physiology at the Second Moscow University (the University for Medicine), and in 1929 she became director of her own scientific research institute, the Institute for Physiology. Her Institute first belonged to the Ministry (Commissariat) for higher education, later it was one of the academic institutes of the Academy of Science of the USSR. She described the aim of her institute in some letters to the neuroscientists Cécile (1875–1962) and Oskar (1870–1959) Vogt in Berlin as follows: she wanted to establish a research program to investigate physiology from the different perspectives of medicine, biology and chemistry. She also wanted to create an international research institute, where scientists from all countries could work and publish together. This aim she could not realize because of the Stalinist policy. For more than 10 years, she and her team worked very successfully. In 1939 Shtern was elected a Full Member of the Academy of Science of the USSR, the first woman scientist of the USSR to be thus honored. Furthermore, in 1944 she became a Member of the newly established Academy of Medical Sciences of the USSR. Already in 1932 she had become a Member of the oldest German Academy of Science (Deutsche Akademie der Naturforscher), the Leopoldina; because of the Nazi regime and the racist policy, she was deleted from the list of members soon after the nomination. After 1945 she was again made a member of the Leopoldina.

Although a member of the Communist Party since 1939, Sthern began to act in politics only when the German troops overtook the Soviet Union in June 1941. Asked for participation in Antifascist Committees, Shtern accepted and became a member of several such committees. In 1941/42 she joined the most important one: the presidium of the Jewish Anti-Fascist Committee (JAC), headed by the famous Yiddish actor Solomon Michoels (Mikhoels) (1890–killed 1948). During World War II (the "great patriotic war", as it was named in the USSR), Shtern worked on war medicine.

In 1948 the tragic part of her life began. Because of the anti-semitic policy of the Soviet State, the Government and leaders of the Communist Party, a campaign was started against "cosmopolitism" which soon led to arrests and deaths, and new discriminations against Jews were established in all spheres of life. Officially stopped in 1953 (after the death of J. V. Stalin), the anti-semitic discrimination policy in reality never ended in the Soviet Union, with fewer arrests after 1953, but with strong barriers against Jews in several professions, including scientific ones.

On January 27th, 1949, Shtern was arrested by the secret service MVD (the secret service in Russia/USSR/Russia was re-named several times: first Cheka, then GPU, later OGPU, then NKVD, after 1945 MVD and MGB, then KGB, now FSB). She was taken to the notorious Lubjanka prison in Moscow, later to the awful prison Lefortovo. She was interrogated, beaten and tortured several times. From 1949 to 1952, she was in prison, together with 14 comrades from the JAC. The trial was planned and prepared at the highest level of the State and under Stalin's leadership, and the end was already fixed: the death penalty. Because some of the prisoners, including Lina Shtern, fought very courageously, the trial was held in secret, from May 8 to July 18,1952 (see Naumov (1994) and Rubinstein (2001)). Although the accused prisoners defended themselves at this trial for several weeks and openly spoke about the tortures and the falsification of the prosecuters, Stalin and his close circle decided to kill them. In August 1952 in Moscow, 13 comrades were killed (one died in the prison). But a miracle happened to Lina Shtern. The dictator Stalin himself struck her name from the list of the death candidates. Until today only rumors are given about the reason for this, the most probable seems to be that Stalin believed in her capacity as a scientist, and hoped she would obtain new medical results that would allow him to live longer.

Thus, Shtern survived the trial and was sent in 1952 by the MVD into exile in Jambul (Central Asia). Fortunately, she was still a Member of the Academy of Science of the USSR, therefore, her salary helped her to survive the difficult living conditions in her exile in a small village. After Stalin's death and the first "rehabilitations" of his victims, Sthern was allowed to come back to Moscow in June 1953. She never got back her own institute, which had been closed in 1948. She only got a position as head of the physiological department at the Institute for Biophysics of the Academy. This important institute became a niche for many political victims of the regime, and it was the cradle of modern molecular biology in the USSR after 1955. In this institute she worked again, and was honored again. But until the end of the USSR, nothing was published about her fate and the fate of the JAC. The obituary of the Academy of Science (1968) was a brief one without any biographical details. Even her successful career in Switzerland was "forgotten". She got an entry in the Great Soviet Encyclopedia, but the years between 1948 and 1955 are "missing", and this "gap" shows knowledgeable readers that "something happened" to her in these years. In 1987, a biography about her was published in Moscow with the same "gap".

Lina S. Shtern worked in two important fields, first biochemistry, especially physiological chemistry, until about 1917. She studied the metabolism, and she studied *in vitro* respiration in special tissues. Furthermore, she worked on the characterization of enzymes involved in substrate metabolism. Between 1904 and 1914, Shtern together with Battelli published about 30 articles on oxidation, mostly in the famous biochemical journal *Biochemische Zeitschrift* of Carl Neuberg (1877–1956). In 1912, Batelli and Shtern published their main results on oxidation and fermentation in a long article (Battelli/Stern (1912) Die Oxydationsfermente, in *Ergebnisse der Physiologie*, **12**, 96–268). Starting in 1917, Shtern studied the effects of certain drugs and organ extracts in organisms. Her new scientific field became the blood–brain barrier. From 1919 to 1923, still in Geneva, she studied the permeability of the blood–brain barrier. Because of her work, she came into close contact with the brain researchers Cécile and Oskar Vogt at the Kaiser Wilhelm Institute for Brain Research in Berlin. Between 1925 and 1929, after she moved to the Soviet Union and had to struggle for her own research institute, she could publish nothing. Then another decade of important research began. Between 1930 and 1940, she carried out new studies on the blood–brain barrier and published, together with Soviet and foreign coauthors, some important papers. During World War II, Shtern worked on war medicine, helping thousands of wounded soldiers; in 1943 she received the Stalin Prize for the practical applications of her medical studies.

Already in 1947, discussions were made against her scientific work and the research program of her institute (see Rapoport (1991), pp. 245–248). She was denounced because she co-operated "too much" with foreigners and employed "too many" Jews in her Institute as well as in the medical journal which she edited. During the years in prison and exile, from 1949 to 1953/55, she had no possibility for any scientific research. Perhaps she was able (and it was allowed) to read some scientific literature during her exile. Less is known about her scientific work in the Institute for Biophysics. She died on March 7, 1968 in Moscow (Soviet Union/USSR).

Lina S. Shtern was one of the first women scientists in Switzerland (as Professor in 1917) and in the USSR (as first female Member of the Academy of Science in 1939). She was one of the founders of modern chemical physiology in the USSR.

Literature

Archive of Cécile and Oskar Vogt, Düsseldorf (address: Medizinische Einrichtungen der Heinrich-Heine-Universität Düsseldorf), correspondence between Lina Shtern (Lina Stern) and Cécile and Oskar Vogt;

Archive of the Academy Leopoldina, Halle/S., related to her membership since 1932.

Archive of the Russian Academy of Science, Moscow.

BSE (Bol'shaja Sovetskaja Encyclopedia), 2oe izd. (2nd edn), (1957) Vol. 48, p. 196, and BSE, 3oe izd. (3rd edn.), (1978) Vol. 29, p. 495.

Lina & the Brain, in Time Magazine, March 3, 1947.

Obituary in Vestnik AN SSSR 5 (1968), p. 118 (very brief, with photo).

Stern, L. (1930) Stern, Lina (Selbstdarstellung (autobiographical sketch)) in *Führende Frauen Europas* (ed E. Kern), Neue Folge, Ernst Reinhardt Verlag, München, pp. 137–140 (with a nice photo); newly published: Conrad/Leuschner (1999), pp. 206–210 + remarks. pp. 270–271 (with great mistakes).

About the JAC and the fate of the 15 comrades:

Hoffer, G. (1999) Lina Stern – Mitglied der sowjetischen Akademie der Wissenschaften (1878–1968) in *Zeit der Heldinnen. Lebensbilder Außergewöhnlicher Jüdischer Frauen* (ed G. Hoffer), dtv, München, pp. 159–184.

Lustiger, A. (1998) *Rotbuch: Stalin und die Juden. Die tragische Geschichte des Jüdischen Antifaschistischen Komitees und der sowjetischen Juden*, Aufbau Verlag, Berlin, especially pp. 371–372.

Lustiger, A. (1994) Die Geschichte des Jüdischen Antifaschistischen Komitees der Sowjetunion. (Nachwort) in *Das Schwarzbuch. Der Genozid an den Sowjetischen Juden* (eds W. Grossman, I. Ehrenburg, German ed. A. Lustiger) Reinbek, Rowohlt, pp. 1093–1101.

Naumov, V. P. (Ed.) (1994) *Nepravednyj sud. Poslednij stalinskij rasstrel. Stenogramma sudebnogo processa nad chlenami Evrejskogo Antifashistskogo Komiteta*, Nauka, Moskva, especially pp. 311–321 and 332–333.

Rubinstein, J. and Naumov, V.P. (Ed.) *Stalin's Secret Pogrom. The Postwar Inquisition of the Jewish Anti-Fascist Committee*, Yale University Press, New Haven and London, (2 photos of Lina Shtern in 1946, the photocopy of the order to arrest Lina Shtern from January 27th, 1949, and a photo of her in the prison), especially pp. 400–416 and 469.

About the scientist Lina Shtern:

Dreifuss, J.J. and Tikhonov, N. (2005), Lina Stern (1878–1968): Physiologin und Biochemikerin, erste Professorin an der Universität Genf und Opfer stalinistischer Prozesse, *Schweizerische Ärztezeitung*, **86** (26),1594–1597.

Grigorian, N.A., Lina Solomonovna Stern (Shtern) in *Jewish Women Encyclopedia*, online.

Jaenicke, L. (2002) Lina Stern (1878–1968). Die biologische Oxydation. Die Schranken und die Erstickung der Forschung, *BIOspektrum*, **8** (4), 374–377.

Ogilvie M. and Harvey, J. (Eds) (2000) Shtern, Lina S. in *The Biographical Dictionary of Women in Science*, Routledge, New York and London, Vol. 2, pp. 1189–1190.

Rapoport, Y. (1991) Lina Stern. Persecution of an Academician, in *The Doctors' Plot of 1953*, Harvard University Press, Cambridge, MA, pp. 234–253 (the book is dedicated to his wife, Sophia Rapoport, who was a student and associate of Lina Shtern; in Russian it was published in Moscow in 1988).

Vein, A.A. (2008) Science and Fate: Lina Stern (1878–1968), A neurophysiologist and biochemist, *Journal of the History of Neuroscience*, **17**, 195–206.

Vogt, A.B. (2007) Lina Shtern (1878–1968), in DSB, N.S. (New Series).

Gertrud Johanna Woker (1878–1968)

Annette B. Vogt

■ Gertrud Woker was a Swiss chemist, one of the first female professors in Switzerland (in Bern), and a very active pacifist and human rights writer. Her scientific fields were research on catalytic processes and problems of biochemistry. She was an active member of the Committee against scientific warfare of the Women's International League for Peace and Freedom.

Gertrud Johanna Woker was born on December 16th, 1878 in Bern. She grew up in a family of academics and clergymen, her father Philip Woker was Professor of the history of religion and history at the University of Bern, her mother was the daughter of a pastor. She was the eldest daughter and had one brother Harald and a sister Elsa. She received a relatively good education, as good as girls could get in her time. Against the resistance of her parents she wanted to study at a university and got the Abitur as an external pupil in 1898. From 1898 until 1900 she attended courses to become a teacher (for girl's schools). From 1900 until 1903 she studied chemistry and biology at the University of Bern where she received the doctoral degree in chemistry, with a thesis on organic chemistry. She was the first Swiss woman to earn a doctoral degree at the Bern University. From 1903 until 1905 she studied at the Berlin University, but only as a guest, because it was not allowed for women to attend the universities in Prussia (only from 1908/09 onwards did it became officially possible). Among others she participated in the courses given by Jacobus Hendrikus van't Hoff (1852–1911) and the biochemist and pharmaceutist Hermann Thoms (1859–1931).

Back from Berlin in 1905, Gertrud Woker had to deal with the problem of finding an academic position which was more difficult because she was a female chemist. Therefore, she first worked in Bern as a teacher at a high school (gymnasium). In 1906 she asked for the procedure to become a Privatdozent at the University of Bern. In January, 1907 she received the venia legendi to teach the history of chemistry and physics there. Her first lecture was called "problems of catalysis research" which described the research program and became her research topic for the next decades.

European Women in Chemistry. Edited by Jan Apotheker and Livia Simon Sarkadi
Copyright © 2011 WILEY-VCH Verlag GmbH & Co. KGaA, Weinheim
ISBN 978-3-527-32956-4

Gertrud Woker (ca. 1928)

From 1911 until her retirement in 1951 Gertrud Woker was the head of the laboratory of physical-chemical biology at the University of Bern. In her laboratory she and her collaborators investigated problems of catalysis and studied biochemical problems. After the publication of her fourth volume in the series *Katalyse* (*Die Rolle der Katalyse in der Analytischen Chemie*) in 1931 in Stuttgart, finally, in 1933, she became a Professor at the University of Bern.

Gertrud Woker became famous and received worldwide acknowledgement because of her contributions to women's, pacifist and human rights movements. She was engaged in these movements throughout her life. Therefore, in 1928, she was asked by the editor, the Belgium journalist, Elga Kern, to write an autobiographical sketch which was published in the book *Leading Women of Europe*. In the 1920s Gertrud Woker published several pamphlets for the "Committee against Scientifc Warfare of the Women's International League for Peace and Freedom" which were distributed in English-, German-, and French-speaking countries. For example, she wrote the pamphlet "A Hell of Poison and Fire", with a drawing by the German artist Käthe Kollwitz (1867–1945), which was distributed in a hundred thousand copies throughout Europe. Her book *Der kommende Gift- und Brandkrieg und seine Auswirkungen gegenüber der Zivilbevölkerung* (The Coming War with Poison Gas) was first published in 1925, and by 1932 nine editions had been published. She dealt with the coming war with poison gas, translated into German the report of the expert committee of the League of Nations, and, in the third chapter, she described the composition and the effects of poison gas, its prohibition and its control. Appeals to scientists were sent out from the Geneva office of the Women's International League for Peace and Freedom. Partly, it became known thanks to the press media. In Switzerland Gertrud Woker ensured that her pamphlets were widely distributed by the central office in Geneva as well as by the Swiss Section of the Women's International League for Peace and Freedom. After 1945 Gertrud Woker was engaged in campaigning against the danger of nuclear wars and for disarmament.

The main research topic of Gertrud Woker was on catalysts. Between 1910 and 1931 she wrote four volumes of *Catalysis*. Furthermore, she published several articles in the series *Chemical and Physical Methods for Biologists and Physiologists*, edited by Emil Abderhalden (1877–1950). In the 1920s she also studied biochemistry problems, for example, the effects of chemical substances in the human body and investigations on the basic elements of plants used as pharmaceutical elements. Her last works, the two volumes *Die Chemie der natürlichen Alkaloide*, were published in 1953 and 1956.

Gertrud Woker trained many – male and female – doctoral students. She also carried out research for chemical and pharmaceutical enterprises in Switzerland. She was one of the pioneers of biochemistry in Switzerland. As an activist in the Swiss and International Women's and Peace movements she campaigned against the dangers of war, of poison warfare and of nuclear weapons. From 1915 until her death she was a Member of the Women's International League for Peace and Freedom (in German the IFFF, Internationale Frauenliga für Frieden und Freiheit). Gertrud Woker died on September 13th, 1968 in Préfargier at the Neuenburg Lake (Neuenburger See).

Literature

Ogilvie, M., Harvey, J. (Eds) (2000) *The Biographical Dictionary of Women in Science. Pioneering Lives from Ancient Times to the Mid-20th Century*, Vol. 2, Routledge New York and London, pp. 1391–1393.

Poggendorff, *Biographisch-Literarisches Handwörterbuch zur Geschichte der exakten (Natur)wissenschaften*, vol. III (1898), vol. IV (1904), vol. V (1926), p. 1387; vol. VI (1937), pp. 2916–2917, VIIa (1956ff.), VIIb (1968ff.), pp. 1062–1063, Leipzig u. a.

Vogt, A. (2007) *Vom Hintereingang zum Hauptportal? Lise Meitner und ihre Kolleginnen an der Berliner Universität und in der Kaiser-Wilhelm-Gesellschaft*, Vol. 17, Franz Steiner Verlag, Pallas&Athene, Stuttgart.

von Leitner, G. (1998) *Wollen wir unsere Hände in Unschuld waschen? Gertrud Woker (1878–1968)* in *Chemikerin & Internationale Frauenliga 1915–1968*, Weidler Buchverlag, Berlin.

Woker, G. *Die Chemie der Natürlichen Alkaloide*, Verlag Enke, Stuttgart, Vol. 1 (1953), Vol. 2 (1956).

Woker, G. (1928) Woker, Gertrud Johanna (Selbstdarstellung (autobiographical sketch)) in *Führende Frauen Europas* (ed E. Kern), Ernst Reinhardt Verlag, München, pp. 138–169 (with photo).

Woker, G. (1925) *Der kommende Gift- und Brandkrieg und seine Auswirkungen gegenüber der Zivilbevölkerung*, Leipzig (6–9 Aufl. Leipzig 1932).

Woker, G. (1913) Die Chemikerin in *Das Frauenbuch*, Frauenberufe und Ausbildungsstätten, Stuttgart, pp. 100–102.

Woker, G. *Katalyse. Handbuch*, Verlag Enke, Stuttgart, Vol. 1 (1910), Vol. 2 (1915), Vol. 3 (1924), Vol. 4 (1931).

Lise Meitner (1878–1968)

Marianne Offereins

■ Although the discovery of nuclear fission, which gave man access to nuclear energy, is mainly associated with the names of Otto Hahn (1879–1968) and Fritz Straßmann (1902–1980), it must not be forgotten that Lise Meitner played an important
part. She was, together with her nephew Otto Frisch (1904–1979), the first to give
physical confirmation of the fission of uranium and to calculate the energy released. Lise Meitner would certainly have played a greater part in the experimental
discovery, if after the *Anschluss* of March 1938 she had not lost her last protection
against the persecution of Jews in Germany.

Elise Meitner was born in Vienna, on October 7, 1878, as the third child of the Jewish lawyer Dr. Philipp Meitner and his wife Hedwig Skovran.

At an early age Lise already showed her interest in physical phenomena, while in
more practical things she could be awkward. Her sisters teased her by saying: *"Es
steht nicht im Physikbüchl!"*[1] After five years at the primary school, followed by
three years at the *Mädchen-Bürgerschule* her education had been completed, further
education was not allowed for girls in those days. However, a few years later, in
1897, girls were admitted to the university. Within two years Lise, by working very
hard, had acquired the curriculum of eight years.

At the University of Vienna, she attended the lectures of Ludwig Boltzmann
(1844–1906), who showed her 'The beauty of Theoretical Physics'. She obtained her
PhD on February 1, 1906 with a doctoral dissertation entitled *Wärmeleitung im in-
homogenen Körper*. In July of that year, her first publication on radioactivity, *Über
Absorption von α- und β-Strahlen* was issued, followed by, *Über die Zerstreuung von
α-Strahlen*.

Because there were no further research opportunities for her in Vienna, she
went to Max Planck (1858–1947) in Berlin, to develop "A fundamental understanding of physics". Lise went to Berlin "for a few semesters". She stayed 31 years!

In Berlin, she was faced with the same prejudices which Sonya Kovalevsky
(1850–1891) had experienced: when Otto Hahn asked her to cooperate with him,

[1] It is not in the physics book.

European Women in Chemistry. Edited by Jan Apotheker and Livia Simon Sarkadi
Copyright © 2011 WILEY-VCH Verlag GmbH & Co. KGaA, Weinheim
ISBN 978-3-527-32956-4

Lise Meitner (© bpk, Berlin).

she had to obtain permission from Professor Emil Fischer (1852–1919). The only women he allowed in his laboratory, were the cleaners. He consented on the condition: *"Wenn sie im Keller bleibt und niemals das Institut betritt, soll es mir recht sein.[2]"* As a laboratory, she was offered an unused carpentry workshop.

During her first years in Berlin Lise lived in a very modest boarding house. She lived mainly on bread and coffee, and she also manifested herself as a heavy smoker.

Between Lise Meitner and Otto Hahn developed a partnership which lasted 30 years, where both worked at the same level. They complemented each other perfectly: while Hahn worked intuitively, Meitner had a more analytical mind, always questioning why.

In physics circles Lise was quickly accepted and she became acquainted with – later – famous scholars including Albert Einstein. Based on his formula $E = mc^2$ Lise would later calculate the energy released by nuclear fission.

In 1906 the duo Hahn–Meitner published six articles on the whole area of beta rays. In 1910, together with Adolf von Baeyer (1835–1917) they published the first recordings of magnetic spectra.

Lise started to work as a paid assistant to Max Planck. Until that time, Meitner and Hahn had their laboratory in the carpentry workshop, which now had become contaminated by radioactivity. Both regularly suffered from headaches and attacks of dizziness.

Approximately one year later Lisa was offered a teaching position in Prague. This recognition led to an offer of a full, permanent position next to Hahn at the insti-

2) If she remains in the basement, I do not care.

tute. Lise decided to remain in Berlin where she could continue her quest for the element which, after release of radiation, led to actinium.

In 1918 Meitner and Hahn discovered the element 91, protactinium, with the symbol Pa. That same year she got her own *Physikalisch- radioaktive Abteilung* in the Kaiser-Wilhelm-Institut. For the first time she earned enough money to set up her own home. She enjoyed *"daß man mit Physik und ähnlichen Dingen Vorhänge erarbeiten kann"* (That one can earn drapes and those kind of things with work).

In 1922 the University of Berlin's Philosophische Facultät awarded Lise Meitner the right to call herself a *professor* and to give lectures. Her opening speech bore the title: *"Die Bedeutung der Radioaktivität für Kosmische Prozesse"*. It was not until March 1, 1926 that she received the official charter in which she was appointed associate professor (Privatdozentin). Her productivity between 1920 and 1933 was enormous. Although Meitner and Hahn each had their own departments, they kept in daily contact.

Lise was fascinated by the studies of Enrico Fermi, who, in 1934, described how he and his researchers bombarded the then heaviest known element – uranium – with neutrons. They thought that during the process new radioactive elements with higher atomic weights emerged, which Fermi called transuranic elements. Meitner, Hahn and his assistant Fritz Straßmann began their own investigations. Lise published in 1935 with Max Delbrück the book *'Der Aufbau der Atomkerne'* and, in 1936, she was nominated for the Nobel Prize.

When Hitler in 1933 rose to power, Lise lost her teaching credentials, and work became more and more impossible for her. From 1936 she could no longer act in public.

Meanwhile, despite all opposition, the research went on. In 1938, just when, after years of research with Otto Hahn, they were about to make the discovery of the century: the fission of uranium, the highlight of their investigations, Lise had to fly from Germany, because she was Jewish and therefore in mortal danger. With the help of Otto Hahn and the Dutch physicists Adriaan Fokker, Peter Debye and Dirk Coster she was able to escape to the Netherlands. On July 14 she arrived, with only a small suitcase, 13 Deutschmarks – a risk because only 10 marks was allowed – and a ring with a large brilliant (diamond), given to her by Hahn, who had inherited it from his mother. Anything else she possessed, she had to leave behind in what was to look like a holiday trip. At 60 years old she had to start again from the beginning, with nothing.

She moved to Copenhagen, where she was received by her nephew Otto Robert Frisch and by Niels Bohr, with whom she was good friends. Bohr tried to persuade her to remain in Denmark, but she moved to Sweden where, at the Nobel Institute in Stockholm, she got an outrageously minor and underpaid job. Especially the attitude of the German government was very difficult for her. She wrote to Hahn: *"...meine Zukunft ist abgeschnitten, soll mir auch noch die Vergangenheit weggenommen werden?"* (My future has been cut off. Will my past be taken away as well?). Daily life in Sweden was not easy for her: the esteemed *Frau Professor* from Berlin lived in a hotel room and could barely obtain a living. Only when, in the middle of 1939, her sister Gusti and her husband, the parents of Otto Robert Frisch, also arrived in Stockholm, could Lise could move to live with them in better accomodation.

During the Christmas of 1938, her nephew Otto Robert Frisch visited her. Later Frisch would call this visit to his aunt: *"den bedeutungsvollsten Besuch meines Lebens"* (The most important visit of my life). By this he meant their elaboration of the first theoretical and physical explanation of nuclear fission by Otto Hahn and Fritz Straßmann in December 1938.

Meitner and Frisch pointed out that a significant proportion of the mass must be converted into energy during the process, they proved that this was indeed the case – the two parts of the shattered uranium nucleus flew away at an incredible speed.

Meitner and Frisch decided to use in their publications for this process not the word 'Zerplatzen' but the word 'fission' because, according to Lise, the division of a unicellular organism in biology is akin to nuclear fission. *"Wenn in dem hochgeladen Urankern (...) durch das eingefangene Neutron die (...) Bewegung des Kerns genügend heftig wird, so kann sich der Kern in die Länge ziehen; es bildet sich eine Art 'Taille', und schließlich erfolgt eine Trennung in zwei ungefähr gleichgroße, leichtere Kerne, die dann wegen ihre gegenseitige Abstoßung mit großer Heftigkeit auseinanderfliegen"*. When in the highly charged uranium nucleus (...) because of the closed in neutron (...) the movement of the core is sufficiently strong, the core can stretch in the length, it forms a kind of 'waist' and finally there is a separation in two roughly equal sized, lighter nuclei, which then fly apart with great violence because of their mutual repulsion". On February 11, 1939, the first publication of Meitner and Frisch on 'nuclear fission' appeared.

Meanwhile, the work of Hahn and Straßmann was also issued. Meitner's name was not mentioned. Possibly Hahn acted under pressure when this happened, but it is a fact, that later neither Hahn nor Straßmann was inclined to mention Meitner's name.

About the war years and her time in Stockholm she wrote in a letter: *"... doch fühle ich mich meistens so einsam, als ob ich in der Wüste lebte"* (Yet I mostly feel as lonesome as if I lived in the desert).

In 1944 Otto Hahn received the message that he had won the Nobel Prize for Physics. In 1946 he received the prize. The work of Meitner was not even mentioned.

On August 6, 1945 the atomic bomb heralded the definitive end of the Second World War. The American scientists could not be reached, the German scientists were stuck in England. Suddenly, the eyes of the world focused on Lise Meitner, who also stood at the cradle of nuclear fission. In America people called her the 'Jewish mother of the atomic bomb'. again and again she emphasized: *"Weder Professor Hahn noch ich selbst haben den leisesten Anteil an der Entwicklung der Atombombe gehabt"* (Neither Professor Hahn nor I took any part in the development of the atomic bomb).

When Karl Herzfeld offered her a guest professorship for the winter season at the Catholic University of Washington, she accepted the invitation. From the wasteland of her exile Lise ended up in a 'madhouse'.

The press descended on her repeatedly with questions about the atomic bomb. The American women journalists, named her woman of the year 1946. In the first three months of her stay in America she received over 500 letters, including many invitations and all sorts of questions, among which was a request for authorization for a film about her life. Otto Frisch writes later in his memoirs, that having a film about her life, would be *"schlimmer als nackt den ganzen Broadway hinunterzu-spazieren"* (worse than walking naked across Broadway).

Though she got several offers to stay in America, she chose to return to Stockholm. In June, 1946, Lise Meitner left the United States on the Queen Mary, laden with honors and having received four honorary doctorates. From Germany came the invitation to take charge of the *Physikalische Abteilung* in Mainz. This invitation she declined as well. In April, 1948, for the first time after all those years, she set foot on German soil to go to Göttingen to attend the memorial service for Max Planck.

In the *Berliner Tagesspiegel* on 10 April, 1953 she is described as '*eine mitfühlende, warmherzige Frau*' (a compassionate warm hearted woman).. The 'great forgetting' has already begun. In many publications, she is considered one of the collaborators of Hahn. Her own work, and the fact that she was in charge of her own physical department, is increasingly pushed into the background.

Meanwhile her working conditions in Stockholm improved, she got more room to work with the necessary equipment, and again she had her own assistants. At the end of 1947 she became a research professor, with the corresponding income. At last her financial worries came to an end.

Otto Frisch had by now moved to Cambridge, followed by his parents. Although Meitner in 1948 took Swedish citizenship – she could keep her Austrian nationality as well – she never felt at home in Sweden.

In 1950 her last publication appeared in *Nature*: '*Spaltung und Schalenmodell der Atomkerne*'.

In total she wrote nearly 150 scientific publications. In 1959 she flew to Berlin for the opening of the *Hahn-Meitner-Institut*. At feminist conferences around the world she was a welcome guest. She also lectured on the effects of the atomic bomb, but the peaceful uses of nuclear energy had her special interest. Politically, she showed great detachment.

In 1960 Lise Meitner moved for the last time in her life, to Cambridge, to be near her nephew Otto and his family. Her health had deteriorated and her hearing got worse, yet she still worked on developments in physics.

In 1966, the team Hahn–Meitner–Straßmann received the American Enrico Fermi Award, as the first non-Americans. It was the first and last time they were honored together for their work. Lise was too old and too sick to accept the prize herself and, therefore, she sent her nephew Otto. On October 27, shortly after noon, she died peacefully. At her wish she was buried next to her brother Walter in the cemetery of St. James Church in Bromley, west of London. On her grave stands, beside her name, and the dates of her birth and death, "A Physicist, who never lost her humanity".

"I realized that this woman, like so many others, was about to disappear from history. The more I researched, the more I realized how unjust that was. (...) But that injustice was then sealed into place by her exclusion from the Nobel Prize, and the subsequent ignorance of journalists and so-called historians who never probed beneath the surface".
Ruth Sime

Literature

Angermayer, E. (1987) Grosse Frauen der Weltgeschichte, in *Tausend Biographien in Wort und Bild*, Neuer Kaiser Verlag – Buch und Welt, Klagenfurt.

van Assche, P.H.M. (1989) De ontdekking van de kernsplijting. in Natuur en Techniek '89, (57), 3.

Bertsch McGrayne, S. (1996) *Nobel Prize Women in Science. Their Lives Struggles and Momentous Discoveries*, Birch Lane Press, New York.

Feyl, R. (1981) *Der Lautlose Aufbruch*. Frauen in der Wissenschaft, Verlag Neues Leben, Berlin.

Jones, L.M. (1990) *Intellectual Contributions of Women to Physics* in *Women of Science,*

Righting the Record (eds. Kass-Simon, G. and P. Farnes), Indiana University Press, Bloomington & Indianapolis.

Kerner, C. (1986) *Lise, Atomphysikerin Die Lebensgeschichte der Lise Meitner*. Belz Verlag, Weinheim/Basel.

Sime, R. L. (1996) *Lise Meitner – a Life in Physics*, University of California Press, Berkeley, L.A./London.

Wertheim, M. (1995) *De broek van Pythagoras. God, fysica en de strijd tussen de seksen. (Pythagoras' Trousers, God, Physics and the Gender Way)*, Anthos, Amsterdam.

Stephanie Horovitz (1887–1942)

Maria Rentetzi

■ Stephanie Horovitz was a chemist who studied in Vienna during the 1910s. She worked closely with Otto Hönigschmid, an atomic weight expert who determined the weight of radium in Vienna and prepared the official substitute of the radium standard in Paris. Horovitz is known for determining, in collaboration with Hönigschmid, the atomic weight of the lead end product of both the uranium and the thorium decay series, providing a persuasive proof of the existence of isotopes. Right after the end of the First World War Horovitz abandoned her research in chemistry and joined the Association for Individual Psychology in Vienna founded by Alfred Adler. In 1924, in collaboration with the psychologist Alice Friedman, Horovitz established a foster home for individual psychological re-education of children and young adults. In 1937 she left Vienna and fled to Warsaw to join her sister. Being a Jew she was not able to escape the Nazi persecution. In 1942 she was sent to Treblinka extermination camp where she was eventually liquidated.

Stephanie Horovitz was born in Warsaw in 1887. Her father Leopold Horovitz belonged to the circle of artists gathered around the court of Emperor Franz Joseph I. Leopold was internationally renowned for his portraits and genre paintings and was able to travel widely in order to carry out portrait commissions. At the World Exhibition of 1873 in Vienna he got the great gold metal for one of his paintings, something that made him more famous in the Austro-Hungarian Monarchy. Leopold reached the apogee of his fame in 1896 when he was asked to paint a portrait of Emperor Franz Joseph I. This and some more lucrative commissions from the Viennese elite enabled him and his entire family – his wife Roza, their three daughters: Zsofia (1877), Janina (1882), Stephanie (1887) and two sons: Georg (1875) and Armin (1880) – to move to Vienna that year. The nine year old Stephanie received home education and completed the university entrance requirements in 1907. That same year she registered at the Philosophical Faculty of the University of Vienna hoping to study chemistry. In 1914 she graduated with a PhD in organic chemistry under the supervision of Guido Goldschmiedt.

In the meanwhile, Horovitz entered the Institute for Radium Research in Vienna in order to help the chemist Otto Hönigschmid in the experimental identific-

European Women in Chemistry. Edited by Jan Apotheker and Livia Simon Sarkadi
Copyright © 2011 WILEY-VCH Verlag GmbH & Co. KGaA, Weinheim
ISBN 978-3-527-32956-4

A family picture with Stephanie Horovitz standing next to her father Leopold Horovitz on the day of her graduation from the University of Vienna in 1913. Behind her stands her mother Roza Horovitz, née London. Her sister Janina Horovitz is the one holding flowers at the far right. Courtesy Piotr Mikucki.

ation of atomic weights of several radioelements with the use of wet chemical techniques. By June 1914 Horovitz and Hönigschmid were working closely together. Hönigschmid informed Lise Meitner "Miss Horovitz and I worked like coolies. On this beautiful Sunday we are still sitting in the laboratory at 6 o'clock". Horovitz and Hönigschmid purified lead from100 kg lead sulfate from the Joachimsthal pitchblende, a time-consuming and meticulous assignment. The atomic weight of radioactive lead was found to be 206.73, lighter than ordinary lead (207.21). On May 23, 1914, Hönigschmid presented their results at a congress of the Bunsen Geselschaft in Leipzig. Conscious of the importance of their work, they immediately sent their article, first to the *Monatshefte für Chemie* instead of the Institute's annual bulletin and shortly afterwards they published a French version in the *Comptes Rendus*.

By the end of the First World War their collaboration was disrupted. Hönigschmid accepted a position at the University of Munich and thus left Vienna. For reasons that are not at all clear, Horovitz left Vienna for a short period of time abandoning her scientific career. According to her family she wished to comfort her mother after her father's death in 1917 and returned to Warsaw. In 1924, however, she returned to Vienna and changed careers. Fascinated by the Adlerian psychology, she joined Alice Friedman in establishing a foster home for children with difficulties in learning. Probably because of the political upheavals, Horovitz left Vienna and moved once again to Warsaw in 1937. Long afterwards, Kasimir Fajans informed Elisabeth Rona that, "Stefanie moved there (to Warsaw) to join her married sister after both her parents had died in Vienna".

When Warsaw was occupied by the Nazis, Horovitz and her sister were able to escape the Jewish Ghetto that Germans built around the Jewish quarter of the city. Beginning in July 1942, the Germans ordered the Jewish leaders to prepare for resettlement to the East and forced the Jews in the Warsaw Ghetto to report "voluntarily" to the Umschlagplatz near the railroad. Horovitz and her sister, being afraid that those who hid them might be persecuted, reported to Umschlagplatz. They were among the thousands Jews who were transported to Treblinka extermination camp and who did not survive.

By the end of the 1910s a considerable number of chemically non-separable pairs or groups of radioelements had begun to accumulate very rapidly. As Frederic Soddy put it colloquially "their atoms have identical outsides but different insides". These elements, identical in their whole chemical character and not separable by any method of chemical analysis, were later called isotopes. In 1913 Soddy succeeded in placing all the known radioelements in the periodic table, despite the fact that there were more of them than the available places. He did so by locating more than one radioelement in the same box, based on the elements' atomic numbers. Even though these elements belonged to different decay series they were chemically inseparable. A promising key to the confirmation of these identical, and at the same time different, substances was a series of comparative atomic weight determinations of isotopic elements. An element available in satisfactory amounts was the two different isotopes of inactive lead end-product of the uranium and thorium decay series.

The experimental work of determining atomic weights was thorny, painstaking, and time consuming. The substances to be weighed must all be isolated in a pure state – a challenging task – and the experimenter should be in a position to determine even the minute quantity of the substance that might get lost during the quantitative experiment. At the time, the world's leading expert in the field was the chemist Theodore Richards, Professor at the University of Harvard and Nobel laureate in 1914 for the accurate determinations of the atomic weight of a large number of chemical elements. Soddy and Kasimir Fajans asked Richards to undertake atomic weight experiments with lead from radioactive sources. Employing the same Harvard method, with the advantage of quartz apparatus, and using lead from Joachimsthal pitchblende, Otto Hönigschmid, another atomic weight expert, at this time in Vienna, repeated the experiments. Carrying out numerous fractionations and crystallizations, he determined the weight of radium and prepared a radium standard, which became the official substitute to the original in Paris.

Performing atomic weight experiments was not a task Hönigschmid could conduct alone, given that already in 1911 he had accepted the directorship of the Laboratory for Inorganic and Analytic Chemistry of the Deutschen Technischen Hochschule in Prague and had become an *Ordinarius* professor there. In January, 1914, he asked Lise Meitner, who was already in Berlin, whether she knew of someone in Vienna qualified to assist him in his atomic weight determination project. Thanks to Meitner's recommendation, Stephanie Horovitz was offered the job. As Hönigschmid wrote to Meitner a few months later, "I am sending you greetings from Miss Horovitz, who does not believe that you still remember her. I have just argued with her about that".

Accounts of Hönigschmid's and Horovitz's joint work, regardless of whether they overlook or overstate her contribution, always present her as his "protégé", a "research student" or simply his "student" who assisted him in determining the atomic weight of radioactive lead. Attempts to emphasize the unfair interpretation of Horovitz's contribution reach the other end of the spectrum by referring to "her results" when they actually site co-published papers. Indeed, to disentangle the politics of collaboration between men and women who work in partnership proves to be a difficult undertaking. Common publications do not reveal who actually took the lead in each project, who was the assistant, and who was assisted. In the case of the Hönigschmid–Horovitz cooperation there is no doubt that he was the mature partner and project leader. He introduced Horovitz to experiments of atomic weight determination and welcomed her both to the Radium Institute and, as it seems, to his laboratory in Prague. In a letter to his friend Max Lembert, Hönigschmid forwarded Horovitz's greetings "With best wishes from Fräulein Doctor Horovitz, the beautiful graduate". Without a doubt Horovitz was more than an able assistant who followed the instructions of her mentor.

This is confirmed by the way Hönigschmid described to Meitner his research project in 1914, emphasizing Horovitz's input in the work. "*We* now isolate lead from pure Joachimsthal pitchblende...*We* hope that in the next two weeks before the holidays *we* will analyze these preparations of lead..." In 1922 in his Nobel lecture Frederick Soddy was the one who acknowledged Horovitz's presence as Hönigschmid's partner, quite the reverse to the account of Richard's team in Harvard. "Simultaneously, work on lead from uranium minerals by T. W. Richards and his students at Harvard, and by Hönigschmid and Mlle. Horovitz, gave values all below the international figure".

As historian Lawrence Badash points out, Hönigschmid and Horovitz offered the most convincing confirmation of the existence of isotopes, at the same time confirming the work that had been done in three other laboratories over the world. They continued to co-publish on the atomic weights of uranium, thorium and ionium until the end of the First World War. In addition, their research showed that ionium was not a separate element but just an isotope of thorium. Horovitz gave up her research in chemistry very early, and began a different career as an individual psychologist in Vienna, joining the Adlerian school of thought.

Literature

Keintzel, B. and Korotin I. (eds) (2002) *Wissenschafterinnen in und aus Österreich*, Böhlau Verlag, Wien.

Rayner-Canham, M. and Rayner-Canham, G. (2000) Stefanie Horovitz, Ellen Gleditsch, Ada Hitchins, and the discovery of isotopes, *Bulletin for the History of Chemistry*, **25** (2), 103–109.

Rayner-Canham, M. and Rayner-Canham, G. (1997) Stefanie Horovitz: A Crucial Role in the Discovery of Isotopes in *A Devotion to Their Science: Pioneer Women in Radioactivity* (eds M. Rayner-Canham and G. Rayner-Canham), McGill-Queen's University Press, Montreal, and Chemical Heritage Foundation, Philadelphia, pp. 192–195.

Rentetzi, M. (2008) *Trafficking Materials and Gendered Experimental Practices: Radium Research in Early Twentieth Century Vienna*, Columbia University Press, New York.

Irén Júlia Götz-Dienes (1889–1941)

Éva Vámos

■ Irén Júlia Götz-Dienes was one of the early researchers in the field of radioactivity in Hungary. She was the second woman to obtain a PhD in chemistry in Hungary, and the first female university lecturer in chemistry in the country. For one year, she worked in Marie Curie's institute. Returning home, she had to switch to other research topics.

Her husband was an outstanding librarian by whom she had three children. Because of the well-known leftist views of the couple they had to flee several times in their lives. Their Odyssey included Vienna, Rumania (Bucharest and Kolozsvár/Cluj) (1923–1928), Berlin (1928–1931), and finally Moscow (1931–1941). There Irén was first appointed by the Nitrogen Research Institute and even became head of department. In 1941 she was sentenced to jail on made-up charges. She was cleared soon but died, in the same year, of typhoid fever that she had contracted in prison.

Irén Júlia Götz was born on April 3, 1889, the daughter of a miller and smallholder in Mosonmagyaróvár, a town near the Austrian border. She studied, as a secondary-school student in one of Budapest's best-renowned institutions for girls, where she finished her studies with distinction. In 1907 she enrolled in the Faculty of Philosophy of the University of Sciences in Budapest, where she studied mathematics, physics, chemistry and also philosophy. From 1908 onwards she was an active member of the Galilei Circle, a group of radical (left-oriented) students, where she had been introduced by László Dienes whom she married in 1913.

She prepared her PhD thesis in radiochemistry: she had to invent a method for measuring radioactive emission. This was not easy as the emission quickly decomposed. She found that by leaving the substance for at least three hours in the apparatus, she could obtain more reliable results as, by that time, the process of decomposition had slowed. She was conferred her title of doctor with the highest distinction 'summa cum laude' and thereupon obtained a scholarship for the academic year 1911/12 to Mme Curie's Paris laboratory where she worked with J. Danysz, a researcher of Polish origin. They even published a short communication on their research into the decomposition products of radium, with special regard to the β-rays of the so-called 'induced activity'. However, Irén fell ill, and had to re-

European Women in Chemistry. Edited by Jan Apotheker and Livia Simon Sarkadi
Copyright © 2011 WILEY-VCH Verlag GmbH & Co. KGaA, Weinheim
ISBN 978-3-527-32956-4

turn home. She had to find a job but could not get one at the university, so had to give up her research on radioactivity. She was employed by the Experimental Station for Animal Physiology and Feeding (ESAPF), from 1915 as an unpaid royal junior chemist. The Station mainly dealt with problems in agricultural chemistry and quality control. In 1919 Irén published a paper on the reasons for the changes in volume observed on mixing liquids (Tamann's law).

In January, 1919 she had the opportunity to give an account, at the Society of (Natural) Sciences, of the results of her theoretical research. This was rewarded by her appointment as lecturer in theoretical chemistry at the University of Sciences. Thus she became the first woman university lecturer in Hungary. After her, a considerable time elapsed before the appointment of the next female lecturers or professors. This came about only in the 1930s.

After the Hungarian Soviet Republic had been defeated, Irén's husband had to leave the country. She could not follow him immediately as she was shortly due to give birth, so she hid in her native town. She was, however, discovered and sent to prison. When released after three months, she joined her husband in Vienna after an adventurous escape. Having no income, they left for Rumania. After a short stay in Bucharest they settled for a while in Kolozsvár/Cluj, where Irén again became a university lecturer, and even published some papers. From the academic year 1922/23 onwards she first lectured in food chemistry, then later became senior assistant professor, and even associate professor at the Institute of Pharmacy. In the year 1927/28 she bore the title of doctor of physical sciences. As Rumania started to become fascist, in 1931 they left for Berlin, where Irén got a job as scientific counsellor with the commercial agency of the Soviet Union. Owing to attacks by the National Socialistic press she was unable to continue, and again they had to seek refuge with their three children, this time in the Soviet Union, in Moscow. Irén was appointed there to a position in the "Nitrogen Research Institute", and even became head of department. In 1938 she was dismissed and could work only as a secondary school teacher. In 1941 she was even tried on the basis of trumped-up charges, and sent to prison. Such changes in people's careers were not rare in the Soviet Union. While imprisoned, she contracted typhoid fever, and, in spite of being released, she died in the same year, at the age of 52. (Her husband was luckier: he survived, returned to Hungary after the war, and first became director of a great public library in Budapest, then head of department at the Faculty of Law of the Budapest University of Sciences until his death in 1953).

Irén Götz was a promising scientific talent in Hungary. She became engaged in one of the most recent and important branches of physics and chemistry as early as at the end of her university studies. At that time she entered Weszelszky's group that was carrying out research into radioactivity. Weszelszky's small laboratory was, at the time, the only one where this branch of science could be studied. Her doctoral theme was focused on finding a method of measurement for the accurate determination of radioactive emission. Her work was based on the utilization of an electrometer developed by Weszelszky. The difficulty encountered by the researchers was that the emission was a very fast process which could not be followed exactly by the instruments at their disposal. It was Irén Götz's idea to leave the substance to be investigated in the instrument for about three hours. By that time the process had slowed down and yielded more reliable data. Irén could not continue her promising work in Hungary as there was no place for her in Weszelszky's laboratory. As described in her biography she had to switch to quite another topic when she got a job with ESAPF. There she had to do much routine work, however, she always found ways to carry out independent research. For example, in 1914 she determined the concentration of the hydroxy ion using a stalagmometer (a device for measuring surface tension). Between 1912 and 1919 she published several papers with Gyula Gróh, later head of the Department of General and Physical Chemistry at Budapest University of Sciences, on topics of animal feeding and physical chemistry. An important paper of hers was that submitted in 1918 to *Zeitschrift für Physikalische Chemie* on the extension of the validity of Tamann's law to the solution of liquids in liquids. During her stay in Rumania she is said to have published several research papers which, however, have not been traced. Her interest in the latest achievements in science never slackened, even when she was deprived of experimental work. Thus, for example, Einstein's theory of relativity had a great impact on her. Einstein was often attacked at this time. She published two papers in defense of his theory. The first paper appeared in *Természettudományi Közlöny* (Scientific Gazette) in 1922, and the second in 1926 in the journal *Korunk* (Our Age), which was started by her husband in Kolozsvár/Cluj in the same year and exists till today. The first three volumes of the journal were edited by Irén Götz.

Irén's role in the science of the epoch merits high appreciation as she joined a branch of science at a time when it was at its beginning in Hungary. So she is to be considered a pioneer of radiochemistry. Her doctoral thesis is worth mentioning as she was one of the first women to achieve a doctorate in chemistry in Hungary. The peak of her scientific career was when she was appointed university lecturer in the Faculty of Philosophy of Budapest University of Sciences. Thus she was also a pioneer for women lecturing at a Hungarian university. Unfortunately, she was only able to enjoy her post at university for a very short time as her appointment was in the epoch of the Hungarian Soviet Republic. After the fall of the Republic she shared the fate of many talented Hungarian scientists of the time, among them George de Hevesy, who later became a Nobel laureate. All lost their jobs, and some even had to leave the country, as did Irén Götz. At this time, when women first had the opportunity of appearing in scientific life – they were admitted to university only after 1895 – most female careers were supported by male scientist members of the family (husbands, fathers or brothers). This did not apply to Irén Götz. She could not rely on any member of her family when entering a scientific career, she could rely only on her talent. Thus she was unique in this respect also.

Literature

Gazda, I. (Ed.) (2004) *Einstein és a Magyarok.* (Einstein and the Hungarians). Akadémiai Kiadó, Budapest, pp. 110–111, 130–132.

Hegedüs-Korach, E. (1983) *The first woman-lecturer in Hungary. Proceedings of the Role of Women in the History of Science, Technology and Medicine in the 19th and 20th Centuries.* Veszprém, August 16–19, 1983.

Hegedüs-Korach, E. (1997) Irén Júlia Götz, in *Magyar Tudóslexikon A-tól Zs-ig.* (Hungarian Encyclopedia from A to Z) Nagy, F. (Editor-in-Chief) BETTER-MTESZ-OMIKK, pp. 340–341.

Palló, G. (1992) *Radioaktivitás és a Kémiai Atomelmélet. Az Anyagszerkezeti Nézetek Válsága a Századelő Magyarországi Kémiájában* (Radioactivity and the Chemi-cal Theory of the Atom. Crisis of the Views on Material Structure in the Chemistry of the Early 20th Century in Hungary), Akadémiai Kiadó.

Palló, G. (2000) A radioaktivitás egy korai kutatója: Götz Irén. (An early researcher of radioactivity: Irén Götz), in *Asszonysorsok a 20. Században* (Women's Fates in the 20th Century). BME Szociológia és Kommunikáció Tanszék; Szociális és Családvédelmi Minisztérium Nőképviseleti Titkársága, Budapest, pp. 33–39.

Radnóti, K. (2008) A Magfizikai Kutatások Hőskora, Női Szemmel – II (Heroic Age of Research in Nuclear Physics, as seen from a female viewpoint). *Fizikai Szemle*, **4**, 150–154.

Erzsébet (Elizabeth) Róna (1890–1981)

Éva Vámos

■ According to a former student of hers, M.D. Marshall Brucer, who wrote a paper "In memoriam" Elizabeth Rona, the radioactive tracer was the discovery of at least four persons of whom Elizabeth was the last one. (If we consider Fajans from Poland, Hevesy from Hungary and Paneth from Vienna as the others, we might say that the discovery of the radioactive tracer was an achievement of the Austro–Hungarian Monarchy). Marelene F. and Geoffrey Rayner-Canham stated that, although she was not involved in any great discovery, she worked with some of the biggest names in the field.

During her long life she worked in six countries including Budapest, Hungary; Karlsruhe and Berlin, Germany; Vienna, Austria; Paris, France; Bornö, Sweden; Washington, D.C., Oak Ridge and Miami, USA. It was she who coined the term 'isotopes' one year after Fajans had discovered them under the name of 'pleiads'.

Elizabeth Róna was born in Budapest, the daughter of Ida Mahler and Samuel Róna, a medical doctor. It was the latter who wanted her to study sciences. However, he thought that the profession of a physician would be too difficult for a woman, thus she enrolled in the Faculty of Philosophy of Budapest University of Sciences, where she studied physics, chemistry and geochemistry. She prepared her doctoral thesis on "Bromine and the monohydric aliphatic alcohols" and was conferred the title in 1912. As soon as she was in her sophomoric year she worked in the Chemical Laboratory of the Veterinary College as a volunteer without payment. Later, she worked at the Chemical Institute No. III. of the University of Sciences.

After finishing her studies she went to work with Fajans in Karlsruhe, and from that time on she devoted all her activities to nuclear chemistry. Shortly before, and during World War I she stayed in Budapest, where she co-operated in chemical courses for students at the University of Sciences. Thus she was the first woman in Hungary to deal with chemistry students. It was here that she first met Hevesy (in 1918), who became interested in her first paper, and thus it came about that they worked together on one of the first applications of the radioactive tracer method. Their joint paper appeared in Germany. Hevesy asked Elizabeth to check a detail in an argument that took place between G.N. Antonoff (Manchester) and F. Soddy

European Women in Chemistry. Edited by Jan Apotheker and Livia Simon Sarkadi
Copyright © 2011 WILEY-VCH Verlag GmbH & Co. KGaA, Weinheim
ISBN 978-3-527-32956-4

Elizabeth Róna (http://www.kfki.hu/physics/historia/
localhost/honap.php?ev=2005&ho=3).

and A. Flecks (Glasgow). It was about a new isotope, today Th-231, discovered by Antonoff, the existence of which could not be verified by the Glasgow scientists. Elizabeth succeeded in confirming the existence of the substance. This work of hers contributed greatly to her reputation.

The controversial situation in Hungary after the War prompted her to leave the country. She went to Berlin, where she worked with Otto Hahn. She had to separate the so-called ionium (today Thorium-230) from uranium ore. In 1924 she was invited to the Vienna Radium Institute, where she had the opportunity to work with Stefan Meyer. Still in the same year the Swedish oceanographer Hans Pettersson presented his samples of sea bottom sediments, which he wanted to have analyzed for radium content, at the Vienna Institute. Entrusted with the work, Elizabeth went to the oceanographic station in Bornö, Sweden. She repeated her visit in the summer for 12 years to investigate the decay chains of uranium, thorium and actinium in oceanographic conditions. These investigations of hers, that revealed the very long half-lives of some of the substances induced her to study radiogeochronology.

At the time she was in Vienna, working with radioactive substances was not considered dangerous. When she asked Stefan Meyer for a gas mask, he simply started laughing. So Elizabeth went and bought two for herself at her own expense. Without them perhaps she would not have lived as long as she did.

During her stay at the Vienna Institute she was sent to Paris to the Curie laboratory. Here she prepared polonium under the guidance of Irene Curie, who told her about their recent discovery of artificial isotopes. The polonium prepared by Elizabeth Róna was given as a gift to the Vienna Institute for research purposes.

Being a Jew, she had to leave Austria after it had become part of Germany in 1938. She first went to Cambridge and then to Oslo (1939), where Otto Hahn spoke to her about nuclear fission as explained by Lisa Meitner. From Oslo she returned to Hungary for a last visit and left for the USA in 1941 to spend the rest of her life there.

Her first position was in the Laboratory of Geophysics at the Carnegie Institute in Washington, D.C.

She participated in the Manhattan project, where her task was to prepare polonium. Her work carried out during World War II was declared as strictly confidential. In 1947 she took up a position at the Argonne National Laboratory, working on uranium reactions. Three years later she became a senior scientist at the Special Training Division of the Oak Ridge Associated Universities (ORAU). From 1954 onwards, when foreign students were admitted, she made good use of her proficiency in a number of European languages.

Her interest in oceanography did not slacken with time. Thus she started a research program in geochronology and geophysics on marine samples.

She retired from ORAU in 1965, at the age of 75. However, she did not stay idle for a minute but took on a position as Professor of Chemistry with the Institute of Marine Sciences of the University of Miami. There she worked on the determination of the composition of sea water using the method of activation analysis.

When she returned from Miami to Oak Ridge for a visit, her former colleagues and friends asked her to write her memoirs. So she did, and the booklet "How it Came about: Radioactivity, Nuclear Physics, Atomic Energy" was published in Oak Ridge in 1978.

Elizabeth Róna was one of the most successful pioneers of radio isotope research. She was much honored in her life, as shown also by the fact that her name was included in an encyclopedia "American Men of Science" published in 1955. Whenever she wanted or was compelled to change the site of her activities, she could find employment with the best renowned institutions of her trade. There she had the opportunity to work with the greatest names, for example, Hevesy in Budapest, Lisa Meitner and Otto Hahn in Berlin-Dahlem, moreover she was invited by Stefan Mayer to join his institute in Vienna.

No encyclopedia dealing with women scientists would be complete without an entry bearing her name. On the whole she can be considered lucky to have lived at a time when a branch of science, of the utmost importance even today, came into being.

Literature

Brucer, M. (1981) In memoriam: Elizabeth Rona (1891–1981) *The Journal of Nuclear Medicine*, **23** (1), 78–79.

Cattell, J. (1980) *American Men of Science. A Biographical Directory*, 9th edn, Vol. I. Physical Sciences, The Science Press, Lancaster, PA – R. R. Bowker Company, New York, p.1637.

Hevesy, G. and Róna, E. (1915) Die lösungsgeschwindigkeit der molekularen schichten. (Solution velocity of molecular layers) *Z. Phys. Chem.*, **89**, 294.

Makra, Z. (1997) Róna Erzsébet, in *Hungarian Scientists' Encyclopedia from A to Z*, BETTER-MTESZ-OMIKK, pp. 684–685.

Palló, G. (1992) Radioaktivitás és a kémiai atomelmélet. (Radioactivity and the chemical theory of the atom. Crisis of the views on material structure in the chemistry of the early 20th century in Hungary.) *Akadémiai Kiadó*.

Palló, G. (1998) Hevesy György. (George de Hevesy) *Akadémiai Kiadó*, Budapest, 69–73.

Radnóti, K. (2008) A magfizikai kutatások hőskora, női szemmel – II. Epizódok a radioaktivitás hazai történetének kezdeteiből. (Heroic age of research of nuclear physics, as seen from a female viewpoint. Episodes from the beginnings of the domestic history of radioactivity) *Fizikai Szemle*, **4**, 150–154.

Rayner-Canham, M.F. and Rayner-Canham, G. (1997) Elizabeth Róna: The Polónium Woman, in *A Devotion to Their Science: Pioneer Women of Radioactivity* (eds Rayner-Canham, M.F. and Rayner-Canham, G.), Chemical Heritage Foundation. McGill–Queen's University Press, Québec, Canada, pp. 209–216.

Rentetzi, M. (2007) *Trafficking Materials and Gender Experimental Practices. Radium Research in the Early 20th Century Vienna*, Columbia University Press, Ch. I, p.58; Ch. II, pp. 69–71.

Róna, E. (1914) Az urán átalakulásairól. (On the transmutations of uranium.) *Mathematikai és Természettudományi Értesítö*, **32**, 350.

Róna, E. (1914) Az urán átalakulásairól. (On the transmutations of uranium). *Magyar Chemikusok Lapja*, **5**, 42.

Róna, E. (1978) *How it Came about: Radioactivity, Nuclear Physics, Atomic Energy*, Oak Ridge Associated Universities.

Gertrud Kornfeld (1891–1955)

Annette B. Vogt

■ Gertrud Kornfeld was the first woman scientist to receive an academic appointment in chemistry at the Berlin University when she obtained the "venia legendi" for physical chemistry to lecture at the University of Berlin and became Privatdozent. She was the first female Privatdozent in chemistry at any University in Germany. She had to flee from Nazi Germany and after several positions she received a job in the United States in an industrial laboratory. Gertrud Kornfeld's life epitomizes both the successes and frustrations of women scientists in academia in the first half of the 20th century.

Gertrud Kornfeld, a daughter of an industrial merchant in Bohemia, was born in Prague on July 25, 1891. Her family belonged to the German-speaking middle-class Jewish families in Prague, respectively, Bohemia. She got an excellent education, first at a German Girl's school then at a boy's gymnasium where she got the matura (the Austrian Abitur, the prerequisite to study at any University). From 1910 until 1915 she studied chemistry, physical chemistry and physics at the German University in Prague. (The famous old Charles University in Prague was divided into a Czech and a German one at the end of the 19th century). In 1915 she finished her study with a doctoral thesis (PhD) at the German University and got a position as an assistant to her doctor father Professor Viktor Rothmund (1870–1927). Because of World War I it became possible for women scientists to get academic positions (as assistants) at several universities.

Because of the political situation and the new circumstances Gertrud Kornfeld left Prague and the Czech Republic in 1918/19 and moved to Germany. As a former assistant of Viktor Rothmund she very soon obtained a position as an assistant to the famous Max Bodenstein (1871–1942) at the Technical College in Hannover. Here she stayed from 1919 until 1923. When Max Bodenstein got the professorship at the Berlin University in 1923 Gertrud Kornfeld followed him as an assistant in his Institute for Physical Chemistry. In 1928 she became a female Privatdozent at the Berlin University in physical chemistry, the first woman in this field here. Furthermore she retained her position as assistant. Gertrud Kornfeld liked to teach and was an advisor for several doctoral theses under the direction of Bodenstein.

European Women in Chemistry. Edited by Jan Apotheker and Livia Simon Sarkadi
Copyright © 2011 WILEY-VCH Verlag GmbH & Co. KGaA, Weinheim
ISBN 978-3-527-32956-4

Gertrud Kornfeld (Vossische Zeitung (Berlin), 1.11.1931)

Because of the Nazi laws Gertrud Kornfeld lost her positions as a Privatdozent as well as an assistant at the Berlin University in the autumn of 1933 and had to go into exile. She went immediately in 1933 to Great Britain. Thanks to the support of the newly established Academic Assistance Council (later the SPSL) Gertrud Kornfeld received a few grants, first at the University of Birmingham, then at the University in Vienna. But as a woman scientist who possessed a relatively high position in Germany she could not find a similar post in exile. Thanks to the American Federation of University Women she got a guest visa for the United States in 1937 to be able to look for an academic position. Finally, she became a researcher at the Eastman Kodak Company in Rochester, New York where she worked until her death on July 4, 1955.

Gertrud Kornfeld was the only woman Privatdozent for chemistry at a German University between 1928 and 1945. She had to change her life and scientific career three times, and she managed to work in scientific research for her whole life, first at universities, and later in the United States in the laboratory of a large industrial company. She first studied problems of organic chemistry, later problems of physical chemistry, and finally she carried out research in photochemistry and kinetics. In the beginning she studied photochemistry from a theoretical perspective, later in the industrial laboratory she studied it from the perspective of applications. Gertrud Kornfeld became the head of a small research group at Eastman Kodak, thus she managed the exile problems successfully.

Literature

Archive of the Berlin University: Phil. Fak. Nr. 1243, pp. 17–39 (thesis); PA Nr. 271 (personal file, 1929–1933, 19 pages).

Archive of the Charles University Prague: Matrikel, thesis documents.

Bio-bibliographical reports in: *Poggendorff: Biographisch-Literarisches Handwörterbuch zur Geschichte der Exakten Naturwissenschaften*, Vol. III–VIIb. Vol. VI (1937), p.1384; Vol. VIIa (1958), p. 880 (Leipzig u. a. 1898).

Biographisches Handbuch der deutschsprachigen Emigration nach 1933 (International biographical dictionary of central European emigrés 1933–1945) 3 Volumes, Vol. II,1 (without date of death) (eds. Röder, W. and Strauss, H. A.) (1980–1983) Saur Verlag, München, p. 651.

List of Displaced German Scholars, London, 1936.

SPSL Archive, Oxford: personal file 218/3, pp. 51–145 (personal file, 1933–1938 + 1955).

Vogt, A. (2007) *Vom Hintereingang zum Hauptportal? Lise Meitner und ihre Kolleginnen an der Berliner Universität und in der Kaiser-Wilhelm-Gesellschaft*, Franz Steiner Verlag, Pallas&Athene, Stuttgart, Vol. 17.

Dorothy Maud Wrinch (1894–1976)

Sally Horrocks

Dorothy Wrinch pursued, in two different countries, a scientific career that included publications in mathematics, philosophy, physics and biochemistry. Initially a successful mathematician with an interest in philosophy, especially of scientific method, her research moved, in the early 1930s, towards biology and chemistry, with the application of her mathematical expertise to structural problems. Her proposed cyclol structure for protein molecules was initially received favorably, but later met with opposition from leading scientists, notably Linus Pauling, as well as being challenged by a growing weight of experimental evidence. These professional struggles were mirrored by turbulence in her private life. Twice married, she spent much of her career in tenuous academic positions that were geographically close to her spouse rather than seeking more secure positions elsewhere.

Dorothy Maud Wrinch was born on 13 September, 1894 in Rosario, Argentina, the elder daughter of Hugh Edward Hart Wrinch, a mechanical engineer and his wife Ada Minnie Souter. She grew up in Surbiton, Surrey where she attended Surbiton High School. In 1913 she was awarded a scholarship to Girton College, Cambridge where she studied mathematics, achieving a first class mark in the part two exams in 1916 and, inspired by Bertrand Russell, moral sciences in which she was awarded a second class mark in part two in 1917. During the academic year 1917/18 she was a research scholar at Girton and in 1918 she won Girton's prestigious Gamble Prize (given to distinguished alumnae) for her work on transfinite numbers. In the same year she was appointed to a lectureship in mathematics at University College, London. After two years in London she returned to Girton as a Yarrow research fellow. In 1921 she was awarded a University of London DSc During this period she was a member of the intellectually impressive circle that surrounded Bertrand Russell and introduced him to one of her Girton friends, Dora Black, who Russell later married. In this milieu she absorbed both feminist and socialist ideas. After her marriage the following year to John William Nicholson, a mathematical physicist and, at the time, a fellow of Balliol College, she remained in Cambridge for a year before moving to Oxford where she became a part-time tutor at Lady Margaret Hall as well as teaching at other Oxford women's colleges.

European Women in Chemistry. Edited by Jan Apotheker and Livia Simon Sarkadi
Copyright © 2011 WILEY-VCH Verlag GmbH & Co. KGaA, Weinheim
ISBN 978-3-527-32956-4

Dorothy Maud Wrinch. (http://www.smith.edu/library/libs/ssc/subjscience.html).

During this period her research was on classical analysis, classical mechanics and mathematical physics, mathematical logic and the theory of scientific method. In 1929 she was awarded a DSc by the University of Oxford, the first woman to achieve this distinction. She was active in the mathematics section of the British Association for the Advancement of Science and served on the International Commission on the Teaching of Mathematics. From the late 1920s her research interests started to diversify. After working briefly on the sociology of child rearing, perhaps in response to her own efforts to combine her professional work with motherhood after the birth of her daughter Pamela in 1927, she moved into theoretical biology, and particularly the application of mathematical techniques to biological problems. The late 1920s were a difficult period for Wrinch, in 1930 she formally separated from her husband, who she described as a 'good mathematician so utterly gone to pieces' (through excessive drinking). The marriage was dissolved in 1938. During the early 1930s Wrinch attended courses in Vienna and Paris to develop her understanding of biology and chemistry and in 1932 was a founder member of the Biotheoretical Gathering, a group of Cambridge biochemists and crystallographers with a particular interest in the structure of proteins and chromosomes. Her earliest publications in her new field proposed possible molecular models for chromosomes. She then moved on to the structure of proteins. Her ideas were initially received favorably, notably at the 1938 Cold Spring Harbor Symposium on Proteins, but she soon became enmeshed in controversy, most notably with Linus Pauling, and resented what she saw as her unfair treatment by a chemical community unwilling to give outsiders a fair hearing.

When World War II broke out she moved to the United States. Employment was not easy to find, perhaps because she had managed to antagonize so many powerful figures during the cyclol controversy before she arrived. She spent her first year there as a visiting fellow in the chemistry department at Johns Hopkins University before securing a joint visiting professorship attached to three colleges in Massachusetts, Amherst, Smith and Mount Holyoke in 1941. This appointment was facilitated by her future husband and a longstanding supporter of her cyclol theory, Otto Charles Glaser, chairman of the biology department and vice-president of Amherst. The couple married in August 1941. Fortunately for Wrinch this was a more successful partnership than her previous marriage. From 1942 she held a special research professorship in physics at Smith, where she worked with Professor Gladys Anslow. In 1943 she became an American citizen. Despite very limited access to funding she managed to continue her research. She supervised a limited number of graduate students and conducted seminars. During the summers she lectured at the Marine Biological Laboratory, Wood's Hole, Massachusetts with which her husband had close links. She was involved with John von Neumann's work on computing at the Institute for Advanced Study in Princeton. She also continued to vigorously defend her theories on protein structure against a growing weight of evidence that challenged it.

Wrinch continued to work at Smith until her retirement in 1971, when she moved to Wood's Hole. She died in 1976, shortly after the death of her daughter Pamela in a fire.

Scientific Work

There were two distinct phases in Wrinch's productive research career. During the first she concentrated her efforts on mathematics but also published on logic and epistemology, influenced by Bertrand Russell and Harold Jeffreys. Her output between 1919 and 1929 comprised 42 publications. Some of these were collaborations with her father and with her first husband. After publishing one sociological text *Retreat from Parenthood* using a pseudonym, Jean Ayling, in 1930 she diverted her attention to theoretical biology, producing a further 150 publications including three books, *Fourier Transforms and Structural Factors* (1946), *Chemical Aspects of the Structure of Small Peptides* (1960) and *Chemical Aspects of Polypeptide Chain Structure and the Cyclol Theory* (1965).

Her focus here was on a search for mathematical solutions to the problems of structure in biological molecules, especially proteins. She is most closely associated with her cyclol theory of protein structure, developed during the late 1930s. Here, she sought a structure that could account for the combination of biological diversity and structural unity that characterized proteins. Her proposed structure required a two-dimensional rather than linear link between the amino acid monomers in proteins, forming sheets rather than chains. Folding the sheets yielded a series of closed octahedra and other solid figures built of amino acid residues. When Wrinch proposed the idea it appeared to fit well with existing experimental data. At the 1938 Cold Spring Harbor Symposium on Proteins she persuaded many participants of the value of her work and gained support from some notable scientists, including Irving Langmuir. This persuaded her to start presenting her model as a theory rather than a working hypothesis. Her ideas were initially well received, in part at least because they appeared to offer a new way forward for protein studies that had been dominated by physical chemists who had long argued that proteins had no definite molecular structure. Wrinch was awarded a five year grant by the Rockefeller Foundation to support her work.

Challenges to her theory started to emerge from among the X-ray crystallography community, particularly workers associated with J.D. Bernal, who challenged her claim that it was supported by X-ray data. W.H. Bragg suggested that X-ray data was in fact inadequate to support a conclusive assessment of any structural theory. These disagreements with British X-ray crystallographers may have contributed to her decision to move to the United States, where she engaged in an acrimonious controversy over protein structure with Linus Pauling, who along with other chemists argued that her structure, despite its geometric elegance, contravened basic principles of their discipline. This confrontation proved very damaging, to her efforts to secure a job or research funding and to her own sense of wellbeing.

Wrinch's association with Glaser and the positions that he was able to help her secure after 1941 brought her the personal stability that had been lacking. In her scientific work she continued to champion her cyclol structure and devoted her research efforts to bolstering it in the face of increasingly marginality to the direction in which debates over the molecular structure of biological compounds were moving. Along with other pioneering scientists whose focus was on proteins, she was increasingly side-lined when the research frontier shifted to nucleic acids in the 1950s. Her most lasting contribution after World War Two was probably the monograph commissioned by the American Society for X-ray and Electron Diffraction, *Fourier Transforms and Structure Factors* (1946) that was used extensively on both sides of the Atlantic.

Dorothy Wrinch was a talented mathematician whose attempts to tackle challenging problems in a new field were not always appreciated by those whose work she disputed. Subsequent assessments of her work have frequently been equally partisan, painting her as misguided and stubborn, or bemoaning the unwillingness of a male scientific establishment to appreciate her genius. These divergent assessments hinge primarily on her continued defense of her cyclol theory in the face of mounting evidence that supported alternative interpretations and her willingness to confront leading experts with more chemical or biological knowledge than she possessed. Sympathizers have suggested that her refusal to reconsider this hypothesis should not detract from the very real contributions she made to the early development of molecular biology through her pioneering methodology, the stimulus her work gave to the study of proteins and to the notion that protein structure should be considered in terms of detailed molecular architecture. Her *Fourier Transforms and Structure Factors* (1946) continued to be regarded as an important tool by crystallographers for many years.

Literature

Abir-Am, P. G. (1987) Synergy or Clash : Disciplinary and Marital Strategies in the Career of Mathematical Biologist Dorothy Wrinch, in *Uneasy Careers and Intimate Lives, Women in Science 1789–1979* (eds P. G; Abir-Am and D. Outram), Rutgers University Press, New Brunswick, NJ, pp. 239–280.

Abir-Am, P. G. (1993) Dorothy Maud Wrinch (1894–1976), in *Women in Chemistry and Physics: A Biobibliographic Sourcebook*, (eds Louise S. Grinstein, Rose K. Rose, and Miriam H. Rafailovich), Greenwood Press, Santa Barbara, pp. 605–612.

Carey, C. W. Jr. (1999) Dorothy Maud Wrinch, *American National Biography* **24** Oxford University Press, New York, pp. 69–71.

Creese, M. R. S. (2004) Dorothy Maud Wrinch (1894–1976), in *Oxford Dictionary of National Biography*, Oxford University Press; online edn, Oct 2007 http://www.oxforddnb.com/view/article/53495, accessed 28 July 2010.

Grinstein, L. S., Rose, R. K., and Rafailovich, M.H. (1993) Dorothy Maud Wrinch, in *Women in Chemistry and Physics : A Biobibliographic Sourcebook*, (eds Louise S. Grinstein, Rose K. Rose, and Miriam H. Rafailovich), Greenwood Press, Westport, CT, pp. 605–612.

Hodgkin, D. C. and Jeffreys H. (1976) Obituary – Dorothy Wrinch, *Nature*, **260**, 564.

Laszlo, P. (1986) Dorothy Wrinch: the mystique of cyclol theory or the story of a mistaken scientific theory, *Molecular Correlates of Biological Concepts* , vol. 34A of *Comprehensive Biochemistry*, Elsevier Science Publishers, ch. 13.

Rayner-Canham, M. and Rayner-Canham, G. (1998) *Women in Chemistry : Their Changing Roles from Alchemical Times to the Mid-Twentieth Century*, American Chemical Society and the Chemical Heritage Foundation, Washington, DC.

http://www.agnesscott.edu/lriddle/women/wrinch.htm

http://www-history.mcs.st-and.ac.uk/Biographies/Wrinch.html

In DNB and American National Biography
http://pubs.acs.org/doi/abs/10.1021/ed064p286.1 article on her by Linus Pauling.

http://pubs.acs.org/doi/abs/10.1021/ed061p890 article to which Pauling is responding.

http://www-history.mcs.st-and.ac.uk/Printref/Wrinch.html list of references – very useful.

http://jchemed.chem.wisc.edu/JCEWWW/features/echemists/Bios/Wrinch.html

Hertha (Herta) Sponer (1895–1968)

Annette B. Vogt

Hertha Sponer was a German-American physicist. She contributed fundamental research to the field of spectroscopy, which became important for physicists and chemists. With her two monographs *Molekülspektren* (published in 1935 and 1936) she contributed to the application of modern quantum mechanics. Hertha Sponer belonged to the small group of German women scientists who were able to establish two academic careers, first at a German University, then later going into exile at an American University. Hertha Sponer was married to the Nobel Prize winner James Franck.

Hertha (also Herta, Hertha Dorothea Elisabeth) Sponer was born on September 1st, 1895 in the town of Neisse in Silesia, the Central European region which is located mostly in present-day Poland. Born into a family of merchants she got family support not only to get a good school education but also to have the possibility to study at a university. All the Sponer children: her brothers, she herself and her younger sister Margot (born in 1898), attended German universities. Hertha Sponer studied physics at the universities in Tübingen and Göttingen. In 1920 she received her doctoral degree at the University of Göttingen with the thesis "Über ultrarote Absorption zweiatomiger Gase", advised by Peter Debye (1884–1966). Already a doctoral student, in 1920, she became an assistant to James Franck (1882–1964), the later Nobel Prize winner, who held an appointment in the Kaiser Wilhelm Institute for Physical Chemistry and Electrochemistry, directed by Fritz Haber (1868–1934) in Berlin-Dahlem. In this year their lifelong collaboration and friendship began. Hertha Sponer became the disciple of James Franck, a close friend of him and his family, and later (in 1946) she became his (second) wife.

In 1921 Hertha Sponer followed James Franck to the University of Göttingen where he became Professor and she held the assistant position. From 1921 until 1933 she held this assistant position, the only one where she was able to receive some payment as a woman scientist. In 1925 Hertha Sponer became Privatdozent at the University of Göttingen after finishing her Habilitation. In the same year she got a fellowship, offered by the Rockefeller Foundation, to do research in the USA, where she worked in 1925 and 1926 in Berkeley. Back in Göttingen she undertook

European Women in Chemistry. Edited by Jan Apotheker and Livia Simon Sarkadi
Copyright © 2011 WILEY-VCH Verlag GmbH & Co. KGaA, Weinheim
ISBN 978-3-527-32956-4

Hertha Sponer (in Maushart (1997))

research, especially on spectroscopy. In 1932 she was nominated as a Professor (außerordentlicher Professor) at the University of Göttingen. Hertha Sponer, Lise Meitner (1878–1968) in 1922 at the Berlin University, and Hedwig Kohn (1887–1964) in 1930 at the Breslau University were the only three women scientists who became Privatdozent for physics at German universities between 1919 and 1945. All three women scientists had to go into exile because of the Nazis.

Because of the Nazi regime in Germany the fruitful and successful work of mathematicians, physicists and chemists in Göttingen was more or less completely destroyed. Like many other German Jewish scientists James Franck had to go into exile. Hertha Sponer, although she was "Aryan", followed him into exile because she did not wanted to live under the Nazi regime. First, from 1934 until 1936 she worked at the University of Oslo. In 1936 she emigrated to the USA where she got an appointment at Duke University in Durham, North Carolina. One of her first assistants here was the physicist Edward Teller (1908–2003).

Hertha Sponer worked at Duke University from 1936 until her retirement in 1966. She became a worldwide acknowledged expert in spectroscopy. Her work was linked not only to physics but also to chemistry. She got her own laboratory at Duke University where she continued her work on near-ultraviolet absorption spectroscopy. One of her colleagues was Hedwig Kohn, after her dramatic escape from Nazi Germany in 1939/40. Hedwig Kohn was a Professor at Wellesley College and, after her retirement in 1952, she joined the laboratory of Hertha Sponer who had supported her escape in 1939.

Hertha Sponer was an extremely successful expert in experimental confirmation by means of spectroscopy. She was one of the pioneers in interdisciplinary research, combining physical methods and chemical problems, developing the spectroscopic investigations as far as possible. She was also a very successful teacher of dozens of doctoral students.

Her sister Margot Sponer (10.2.1898–27.4.1945(murdered)), a scholar in Romanistic studies, a teacher at the Berlin University, where she received her doctoral degree in 1935, and an interpreter of Spanish, took part in resistance activities against the Nazis. Therefore, she was killed by the Gestapo in the very last days of World War II. On April 27, 1945 she was shot in Berlin-Wilmersdorf, shortly before the Red Army reached and liberated this area.

In 1946, Hertha Sponer and her friend and colleague James Frank were married. After his death, and after her retirement, she went to Germany where she had relatives. She died on February 17th, 1968 in the little town of Ilten near Hannover.

Since 2002 the German Physical Society (Deutsche Physikalische Gesellschaft) awards every year the "Hertha Sponer Prize" to a young woman physicist, to encourage women scientists.

Literature

Archive of the Kaiser-Wilhelm-/Max-Planck-Gesellschaft, Berlin.

Archive of the University of Göttingen.

Biographisches Handbuch der deutschsprachigen Emigration nach 1933 (International Biographical Dictionary of Central European Emigrés 1933–1945) (1983), Vol. II,1 (without date of death) (eds. Röder, W. and Strauss, H. A.) (1980–1983) Saur Verlag, München.

Maushart, M.A. *"Um mich nicht zu vergessen": Hertha Sponer – ein Frauenleben für die Physik im 20. Jahrhundert*, GNT Verlag, Bassum/Stuttgart.

Ogilvie, M. and Harvey, J. (Eds) (2000) *The Biographical Dictionary of Women in Science. Pioneering Lives from Ancient Times to the Mid-20th Century*, Routledge, New York and London, Vol. 2, pp. 1220–1221.

Poggendorff, *Biographisch-Literarisches Handwörterbuch zur Geschichte der exakten (Natur)wissenschaften* Vol. III (1898), IV (1904), V (1926), VI (1937), p. 2515; VIIa (1956ff.), pp. 465–466; VIIb (1968ff.) Leipzig u. a.

Tobies, R. (1996) Physikerinnen und spektroskopische Forschungen: Hertha Sponer (1895–1968) in *Geschlechterverhältnisse in Medizin, Naturwissenschaft und Technik* (eds C. Meinel, and M. Renneberg), GNT Verlag, Bassum/Stuttgart, pp. 89–97.

Vogt, A. (2008) *Wissenschaftlerinnen in Kaiser-Wilhelm-Instituten. A-Z. 2. erw. Aufl.*, (Veröffentlichungen aus dem Archiv zur Geschichte der Max-Planck-Gesellschaft, Bd. 12), Berlin, pp. 176–177.

Vogt, A. (2007) *Vom Hintereingang zum Hauptportal? Lise Meitner und ihre Kolleginnen an der Berliner Universität und in der Kaiser-Wilhelm-Gesellschaft*. Franz Steiner Verlag, Pallas&Athene, Stuttgart, Vol. 17.

Gerty Theresa Cori (1896–1957)

Marianne Offereins

Gerty Cori was the third woman – after Marie Curie and Irène Joliot-Curie – to win a Nobel Prize. She was the first woman to win the Nobel Prize for Medicine. She received the prize in 1947 with her husband for the discovery of the catalytic conversion of glycogen.

On August 15, 1896 Gerty Theresa Radnitz was born in Prague, as the eldest daughter of Martha and Otto Radnitz Neustadt, who was a chemist and director of a number of sugar refineries.

Gerty had two younger sisters, Lotte and Hilda.

As was customary for people of her social circle in those days, Gerty was tutored at home until she was ten years old. After that she visited a girls finishing school, where she graduated in 1912. This diploma did not give access to the university, where at first she wanted to study chemistry; at the age of 16, however, she decided

Gerty Theresa Cori
(Bernhard Becker Medical Library).

European Women in Chemistry. Edited by Jan Apotheker and Livia Simon Sarkadi
Copyright © 2011 WILEY-VCH Verlag GmbH & Co. KGaA, Weinheim
ISBN 978-3-527-32956-4

that she would study Medicine. Therefore she went to the Gymnasium where, two years later, she passed her final exam.

In 1914 Gerty Cori registered as a student at the medical faculty of the University of Prague, because she had discovered that biochemistry could be studied there. During her first year Gerty met Carl Ferdinand Cori, who had started his studies in the same year. More than 50 years later he wrote that, right from the beginning, he was very impressed by her "charm, intelligence, vitality and sense of humor". From that moment on they studied together until Carl was drafted to serve as a 'Sanitätsoffizier' in the Austrian army during the First World War. In 1918 he returned to Prague, where he resumed his studies. The collaboration of Gerty and Carl restarted and would go on until her death in 1957. In 1920 Gerty and Carl obtained their medical degrees.

Because Carl wanted to concentrate on the scientific side of medicine, he moved to Vienna where he divided his time between the University Clinic for Internal Medicine and the Institute of Pharmacology.

Six months later Gerty followed him, and on August 5, 1920 Carl Cori and Gerty Radnitz were married.

After her marriage Gerty could continue her career. She had a position in Vienna as an assistant at the Karolinen Children's Hospital, where she could specialize as a pediatrician. However, like her husband, Gerty was more interested in the purely fundamental scientific work.

Due to the unstable situation in Europe, Carl and Gerty decided to leave Europe at any cost. Carl got an appointment at the *New York State Institute for the Study of Malignant Diseases* in Buffalo (now *Roswell Park Memorial Institute*) where patients were treated with X-rays and radium radiation. In 1922 he departed for the 'Neue Welt'. Again Gerty followed her husband after six months.

During the next 25 years she would have to accept lower positions, with lower income, and occasionally even without pay. She received an appointment as a radiographer at the same institute as Carl. The work was mostly routine and Gerty spent much of her remaining time helping her husband with his research. The director of the institute objected fiercely, so from then on she did her research for Carl more discreetly. From the beginning of their relationship the Coris were perfectly adjusted to each other: outside their work they had the same hobbies and interests. Together they discussed their experiments which led them to excellent results. It happened often that in a conversation one of the two began a sentence, while the other person finished it. Acquaintances said about this collaboration: it looks like together they use one brain. The precision of their work became their characteristic. According to her biographer J. Larner, Gerty was: "Undoubtedly primarily responsible for the development of the quantitative analytical methodology".

The years in Buffalo were important for the Coris. Here they had the opportunity to adjust to American life and to find their own niche of research. They produced nearly 100 publications. During the 1920s the Coris carried out research on the use of glucose in the muscles. By 1929 they were able to explain how mammals get their energy for heavy muscular exercise. According to their theory, glucose moves in a

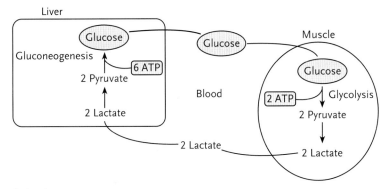

Cori cycle.

cycle from the muscle to the liver and back to the muscle again. They called this the *cycle of carbohydrates.* Everybody else called it the 'Cori-cycle'.

In 1928 the Coris became naturalized U.S. citizens.

When Carl Cori in 1931 was appointed professor at Washington University in St. Louis, Missouri, Gerty followed him as research assistant in pharmacology. Here too she did most of her work in a lowly position. Her salary was 1500 dollars, twenty percent of Carl's pay, but it was far better than any offer she could get elsewhere. In 1936 the Cori couple isolated from the muscles of a frog Glucose I-phosphate (now also known as the 'Cori ester'), until then an unknown intermediate in the construction of glucose.

In the meantime, it had become clear that enzymes are very important in this process, so the Coris changed the direction of their research toward enzymology. This led to the discovery of the enzyme phosphorylase, which breaks glycogen down into the Cori ester. A few years later their laboratory succeeded in crystallizing phosphorylase; after that they discovered one enzyme after another. Their work proved to be of enormous influence on the research into and the treatment of diabetes and other metabolic diseases.

Cori ester.

In August 1936 Gerty had to interrupt her research for a private situation. Her son, Carl Thomas was born. As the contractions began, she went on working in the laboratory, from there she was taken straight to the maternity ward, where a few hours later she gave birth to her son. Three days later she was back in her laboratory.

In 1943 (or 1944, sources differ on that) at last Gerty became Associate Professor of Biochemistry at Washington University and finally, in 1947, she received an appointment as Professor.

For their discovery of the catalytic conversion of glycogen in 1947 the couple was awarded the Nobel Prize, together with the Argentine Alberto Bernardo Houssay. Gerty Cori was thus the first woman awarded the Nobel Prize for Medicine, in addition she was the first American woman ever to get a Nobel Prize. A few weeks before she went on a trip to the awards ceremony, Gerty's physician had to tell her that she suffered from a fatal type of anemia, now known as myelofibrosis. For the rest of her life she would be dependent on blood transfusions.

During the last decade of her life, despite her serious illness, she worked on and discovered that a genetic defect is the cause of enzymatically pathological glycogen accumulation in children.

On October 26, 1957 Gerty Cori died from a kidney ailment. She was 61 years old.

Literature

Fölsing, U. (1191) *Nobel-Frauen. Naturwissenschaftlerinnen im Portraet*, Verlag C.H. Beck Munich.

Kerners, C. (1991) *Nicht nur Frauen Marie Curie ... Frauen die den Nobel Prize Bekamen*, Beltz Verlag, Weinheim und Basel.

McGrayne, B.S. (1996) *Nobel Prize Women in Science. Their Lives, Struggles and Momentous Discoveries*, Birch Lane Publishers, New York.

Pycior, H.M., Slack, N.G., and Abir-Am, P.G. *Creative Couples in the Sciences* (eds.) Rutgers University Press, New Brunswick, New Jersey.

Strohmeier, R. (1998) *Lexicon der Naturwissenschaftlerinnen und naturkundigen und Frauen Europas. Von der Antike bis zum 20. Jahrhundert*. Harri Deutsch Verlag, Thun und Frankfurt am Main.

Ida Noddack-Tacke (1896–1978)

Marianne Offereins

■ Ida Noddack-Tacke was, together with her husband, Walter Noddack, one of the discoverers of the element rhenium. Based on the fact that there was an open space in the Periodic Table, they calculated the characteristics of rhenium and, after years of research, isolated the element in 1925.

On February 25, 1896, Ida Tacke was born as the third daughter in the family of varnish manufacturer Albert Tacke and his wife Hedwig Danner in Lackhausen near Wesel in the Rheinland.

At the age of 16 she was accepted at the St.-Ursula-Gymnasium in Aachen. After passing her final exam at this school Ida studied chemistry at the Technische Hochschule in Berlin. In 1919 she became Diplom-Ingenieur.

She attained her Doctor's degree in 1921 at the Laboratorium für Fettforschung of the same Technische Hochschule with a thesis: *Die anhydride höherer aliphatischer Fettsäuren* (Higher aliphatic fatty acid anhydrides). After that she found positions in Berlin at the AEG and Siemens & Halske factories. Here she was the first woman in German industrial research.

During this work Ida became a specialist in X-ray spectroscopy. The determination of trace elements, investigation of the origin and the concentration of elements in nature, were the areas in which she specialized. Furthermore, she investigated the quantitative determination of special elements in minerals and meteorites, for which she produced new separation and enriching methods.

In 1922 Ida Tacke was appointed Gastwissenschaftlerin (visiting scientist) at the Physikalisch-Technische Reichsanstalt in Berlin.

There she started, together with the head of the laboratory, chemist 'Regierungsrat' Dr. Walter Noddack, research for elements which could fill in a number of gaps that still remained in the Periodic Table: the places 43 and 75 in Group VII under manganese. These two elements were indicated as 'Ekamanganese'. These elements were very rare and occured in two forms: pure, about 1% in platinum ore and ten times as rare in metal oxides like columbite (niobite).

Ida Tacke and Walter Noddack gave very precise predictions about the atomic masses, melting points, even about the colors and the forms of the crystals and the

European Women in Chemistry. Edited by Jan Apotheker and Livia Simon Sarkadi
Copyright © 2011 WILEY-VCH Verlag GmbH & Co. KGaA, Weinheim
ISBN 978-3-527-32956-4

Ida Noddack (Women at Work Museum).

chemical behavior of both ekamanganeses. Based on that knowledge they chose the chemical methods to isolate the elements they were looking for.

Finally, they succeeded in obtaining 1 mg of the element which was lost during further chemical research. It was in the middle of the Great Depression after the collapse of the Stock Market, inflation was sky high, new platinum was far too expensive.

So they worked up 1 kg of columbite, from which, after far more difficulties than with the platinum ore, they obtained 1 mg of a new element.

With this Ida went to the firm Siemens & Halske, where she investigated the sample with the new X-ray spectroscope, together with Otto Berg. Finally, on July 11, they could observe the characteristic spectral lines of the elements 43 and 75.

In 1925, they published a paper (*Zwei neue Elemente der Mangangruppe, Chemischer Teil*) claiming to have done so. The element 75 they called rhenium, from *Rhenus*, Latin for the Rhine, the area where Ida came from; the other element, 43, masurium, was named after Masuria in former East Prussia, where Walter Noddack originated from. Only the discovery of rhenium was confirmed.

They were unable to isolate element 43 and their results were not reproducible. Now known as technetium, element 43 has never been found in nature. In 1937 it was produced artificially. Therefore during an IUPAC-conference in Amsterdam (1949) it was called technetium. It is significant, however, that Otto Hahn and Fritz Strassmann in their publications extensively refer to the element *masurium*.

In the following years they processed, with financial support from Siemens & Halske, 660 kg of molybdenite. In 1928 they had their first gram of pure rhenium.

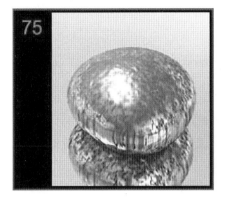

Rhenium.

The research had cost 50 000 DM. Ida Tacke's part in this investigation is indisputable: she worked on the chemical research with Walter Noddack, and the spectral analysis with Otto Berg, however, in both cases Ida was the central scientist of the investigation.

In 1926 the collaboration between Ida Tacke and Walter Noddack extended even further. From that moment on they shared the rest of their lives as well: on May 20, Walter and his 'Idchen' were married. They did not have any children, which – as sources say – must have been quite difficult for them.

Ida Noddack's career became strongly intertwined with her husband's and together they published about a hundred scientific papers.

In 1934, after Enrico Fermi and his group in Italy had bombarded uranium with neutrons, and had concluded that they had produced transuranium elements, artificial elements heavier than uranium, Ida Noddack suggested he had split uranium atoms into isotopes of known elements rather than added to uranium atoms to produce heavier, unknown elements. At that time an utterly presumptuous thought.

She presented her point of view in September, 1934 in an article with the title: '*Über das Element 93*', in the *Zeitschrift für Angewandte Chemie*. In retrospect it proved to be a brilliant suggestion.

Not thwarted by the perceptions of the structure of the atomic nucleus of that time, she advised Fermi to first eliminate all known elements – and not just the elements between $Z = 82$ and $Z = 92$ – before claiming to have found new elements with $Z = 93$, $Z = 94$ and so on. She wrote: "One can imagine that by bombarding heavy nuclei with neutrons these nuclei may disintegrate into more big pieces, which are isotopes of known elements, but not neighbors of the irradiated elements". This remark could have put the whole radiochemical community on the right track, but fellow chemist Otto Hahn later commented in his autobiography "Her suggestion was so out of line with the then-accepted ideas about the atomic nucleus that it was never seriously discussed". In 1939, after much research by many scientists, Otto Hahn, Fritz Strassmann and Lise Meitner discovered that Noddack had been right. They named the process nuclear fission.

In 1938 Fermi was awarded the Nobel Prize for his mistake.

At the same time the couple Noddack–Tacke published the chemical and physical equations of the transuranic elements to the element 118 which were until then unknown. They also succeeded in detecting a natural radioactivity in platinum ore.

After the war Walter Noddack initiated a course in geology at the Katholische Hochschule in Bamberg.

Until Walter Noddack's death, December 7, 1960, Ida and he worked together at the Geochemical Institute in Bamberg.

The Noddack couple was very popular among their students and fellow workers. They treated them as equals (no one could ever outdrink them – the contrary occurred several times).

Until 1968, when she retired at the age of 72, Ida kept on working at the institute. After her retirement Ida went on with her research. Among others she investigated the chemical solubility of kidney stones.

Ida Noddack-Tacke died on September 24, 1978 in Bad Neuenahr where she spent the last years of her life. In accordance with her last will she was cremated. Her urn was buried in the grave of Walter Noddack in Bamberg.

Special thanks to Prof. Dr. ir. Pieter van Assche, Leuven, and Dr. Renate Strohmeier, Frankfurt.

Literature

Angermeyer, Dr. E. (1987) *Grosse Frauen der Weltgeschichte. Tausend Biographien in Wort und Bild*. Neuer Kaiser Verlag – Buch und Welt, Klagenfurt.

Assche, P.H.M. van (1989) *De ontdekking van de Kernsplijting, een kettingreactie van gemiste kansen*. Natuur en Techniek '89, 57. 3, 170–183.

Assche, P.H.M. van (1988) *Ignored priorities: first fission fragment (1925) and first mention of fission (1934)*. In: *Nuclear Europe* 6-7/ 1988.

Assche, P.H.M. van (1988) *The ignored discovery of the element Z=43* in: *Nuclear Physics* **A480** 205–214.

Kass-Simon, G., & Farnes, P. (1990) *Women of science; righting the record*. Indiana University Press, Indiana.

Kerner, C. (1986) *Lise, Atompysikerin*. Beltz Verlag, Weinheim & Basel.

Noddack, W., Tacke, I. und Berg, O. (1988) *Die Ekamangane. The Ekamanganese elements*, Translated by G. Michiels and P. van Assche. Studiecentrum voor Kernenergie, Mol..

Pflaum, R. (1989) *Grand Obsession. Madame Curie and her world*. Doubleday, New York.

Sime, R. (1996) *Lise Meitner, a life in physics*. University of California Press, Berkeley, Los Angeles, London.

Tilgner, H.G. (1999) *Forschen, Suche und Sucht. Kein Nobelpreis für das Deutsche Forscherehepaar das Rhenium entdeckt hat. Eine Biographie von Walter Noddack (1893–1960) und Ida Noddack-Tacke (1896–1978)*. Hans Georg Tilgner, Books on Demand GmbH, Mülheim an der Ruhr.

Ilona Kelp-Kabay (1897–1970)

Éva Vámos, István Próder, and Katalin Nyári-Varga

Ilona Kelp's most important research was carried out together with her husband János Kabay. It was aimed at developing a method for the large-scale production of morphine from the green poppy plant. The patented process was implemented at a factory founded by the family in 1927 in Tiszavasvári, a village in North-East Hungary. The factory "Alkaloida" is still in operation.

Ilona Kelp was the third woman to obtain a doctor's degree in chemistry in Hungary, and the first – and for the time being – the only one, whose portrait in oil is shown in the permanent exhibition of the Chemistry Museum of the Hungarian Museum for Science, Technology and Transport in Várpalota. She gave up her career as a chemist when her husband died at the age of 39 in 1936. After the factory had been nationalized in 1948, she left for Austria, and ultimately settled in Australia with her two children.

Ilona Kelp was born in Kassa (today Košice, Slovakia) on 25 September, 1897. Her youth and studies were poverty-ridden as her father, a lieutenant colonel of the Hungarian army, died at a young age. At the age of 15 Ilona, the youngest of three children, had to make her own provision for her studies. So she took on a job as clerk at the Ministry of Agriculture and could not attend secondary school. However, she passed the final exam with distinction as a private student. The family having moved to Budapest, she enrolled in the Faculty of Philosophy of the University of Sciences as a student of chemistry. There she had the opportunity of studying with the best Hungarian professors of the time. It is perhaps interesting to mention that one of them, the famous analytical chemist Lajos Winkler, published a paper in which he called the burning of poppy straw as fuel sheer madness as this step destroyed tons of the alkaloid morphine contained in all the parts of the poppy plant. Morphine and its derivative codeine were, and still are, used in medicine as excellent pain-killers. Ilona's talent and assiduity were so highly appreciated that she was even allowed to attend Professor Géza Zemplén's course in organic chemistry at the Technical University, where – as a rule – female students were not accepted.

European Women in Chemistry. Edited by Jan Apotheker and Livia Simon Sarkadi
Copyright © 2011 WILEY-VCH Verlag GmbH & Co. KGaA, Weinheim
ISBN 978-3-527-32956-4

Ilona Kelp (http://magyarmuzeum.org/
uploaded/images/20080313-131642_0.jpg).

Ilona Kelp prepared her doctoral thesis on the diffusion velocity of iodine in different solvents, and obtained her title of doctor of chemistry with distinction. Soon thereafter, in 1924, she took on a job as researcher at the Experimental Station for Medicinal Herbs, where she started dealing with the analysis of the poppy plant's morphine content. It was pure chance that a young pharmacist, János Kabay (1896–1936), took on a job at the same institution in the same year. He was engaged in research on the morphine content of the green poppy plant. The two young scientists met, fell in love, and got married in 1925. In December of the same year the young couple moved to Ilona Kelp's flat in Budapest. This was a good solution for them as in March, 1926 their first child – a son – was born, and Ilona's mother helped care for the baby while the parents were at work.

Still in 1925, János Kabay patented his process for the large scale production of morphine from the green poppy plant. Although Ilona Kelp's analytical work was of great merit in her husband's achievement, she did not want to have her name mentioned in the patent as co-inventor.

As the pharmaceutical industry did not support Kabay's work, the couple quit their jobs with the Station, and moved to the young husband's birth place in Eastern Hungary (Büdszentmihály, today Tiszavasvári). There the family pooled their resources and founded the share company "Alkaloida Vegyészeti Gyár Rt." (Alkaloida Chemical Factory SC). Production started in Tiszavasvári in 1927. In that year the weather was particularly changeable: owing to the extreme heat, fermentation started in the crude poppy extract, which caused great losses in yield. The whole morphine production amounted to only 1900 g, which caused disappointment among the shareholders.

The factory sent their product to the Faculty of Pharmacy of the University for their opinion. They confirmed that the product met the requirements of many European Pharmacopoeias and of that of the USA. The University appreciated the large scale results and expressed the opinion that it would be desirable if the factory's conditions permitted the suitable utilization of this "special Hungarian invention" in the country. This favorable opinion helped the factory to obtain a state loan. This, however, did not cover the expenses of development. The factory was rescued by János's brother Peter selling his pharmacy. Still, the problem of transporting the poppy plant had to be solved to ensure undisturbed production. The problem was solved by János Kabay's new invention: a machine which allowed extraction of the morphine from the plant on the spot. In 1928 the yearly production rose to 10 kg of crude morphine. In January, 1929 the shareholders said that it would be desirable if the factory extracted 100 kg of crude morphine in a year. This entailed enlarging the premises and purchasing new equipment, which brought the company near to bankruptcy. However, it survived.

In 1931 Kabay patented a process for obtaining morphine from dried poppy straw, a waste of poppy processing until then unutilized. Later the process was extended to dried poppy-heads as their morphine content was much higher than that of the straw. In all this Ilona Kelp's analyses and her systematic arrangement of her husband's work were of great help.

The difficulties of production, the distrust of the creditors and the lack of harmony within the family – mainly the jealousy of his brother Peter, the pharmacist – drove Kabay to change places with his brother: he left him to run the production, and moved to Budapest with his family, to the central office of the company in the capital. Here the couple hired a laboratory at their former institute where they could continue their research together as in the early happy times.

In the meantime production was extended to codeine and other derivatives of morphine. Here again, Ilona's analytical skill was much needed in developing methods for the analysis of the new products.

On one occasion, during Kabay's stay in the capital in 1933, the permanent Head of the Ministry of Health invited him to a dinner organized in his honor. High officials of the Ministry, the Police and the Physician's Association were present. The host praised the merits of the Kabay couple for nearly an hour, and János was near to tears on hearing how highly Ilona was appreciated by the speaker.

Finally, in 1934, things seemed to improve for the company and for János Kabay as well: a factory was being built in Poland based upon his patent and process, and after his designs. He himself had to supervise the building progress, and the first steps of production. This meant that he had to be absent from the company for shorter or longer periods, several times. During his absence Ilona, the only person he always could absolutely rely upon, directed the works as she was well acquainted with every step of the production.

Amidst all this work, in 1936, János Kabay died unexpectedly – probably as a consequence of medical malpractice – at the age of 39. Ilona Kelp carried on as president of "Alkaloida" for a short time. After her husband's death she finished the series of investigations started upon the request of the League of Nations' Commission on Narcotic Drugs. She sent her account of the results to the seat of the League in Geneva. Her paper was published in the Bulletin of the Hungarian Society for (Natural) Sciences in 1936. Thereafter, however, she retired, and stopped dealing with chemistry and pharmacy for good.

When "Alkaloida" was nationalized in 1948, she left Hungary for Austria with her two children. From 1950 till her death she lived in Sydney, Australia. Today, she is commemorated by a pharmacy in Tiszavasvári bearing her name. The pharmacy is located in János Kabay street.

In the history of science we often encounter men of great achievements with wives that were companions, not only in life but also in their profession. Ilona Kelp was such a companion to the outstanding inventor János Kabay. She herself being of excellent qualities in her profession always wanted to remain in the shadows. However, the unique process of morphine production from the dried poppy plant, and the factory implementing the process as invented and carried out by her husband, could not have come into being without her contribution, furnishing all the analytical work needed in the production and in the quality control of the product. It was her scientist's eye and brain that found, by assiduously systemizing the results of raw material analysis, that the morphine content of the poppy plant varied quite considerably with the region of origin. As this could be due to the soil or the seed material only, it was decided to supply all the poppy growers with seed material from the region in which the plants of the highest morphine content were produced. Thereafter these differences, which made uniform production difficult, could be eliminated. Thus analytical work, often rated as of secondary importance, might be helpful in improving technological results.

Another important contribution to the history of science is Ilona Kelp's diary that helped, among others, her son John J. Kabay when writing a book about his father's life and struggles in creating a new and important production process for drugs indispensable in medicine.

Literature

Hosztafy, S. (1997) 100 éve született Kabay János, a magyar morfingyártás megalapítója. (János Kabay, the founder of Hungarian morphine manufacture, was born 100 years ago.) *Élet és Tudomány*.

Kabay, J.J. (1992) *Kabay János Magyar Feltaláló Elete*. (Life of the Hungarian Inventor János Kabay). Alkaloida Vegyészeti Gyár Részvénytársaság, Tiszavasvári.

Próder, I. and Varga-Nyári, K. (1997) Arckép avató ünnepségek a Magyar Vegyészeti Múzeumban. Náray-Szabó István és Kelp Ilona arcképének leleplezése. (Inauguration of two portraits in the Hungarian Museum of Chemistry: The inauguration of the portrait of István Náray-Szabó and that of Ilona Kelp (1897–1970)) *Magyar Kémikusok Lapja*, **52** (12).

Varga-Nyári, K. Kelp Ilona (1897–1970). Társ, nemcsak az életben. (Ilona Kelp (1897–1970). Companion, not only in life)

Irène Joliot-Curie (1897–1956)

Renate Strohmeier

■ Irène Joliot was the second woman, after her mother Marie Curie, to win the Nobel Prize in Chemistry

In 1937 Irène Joliot-Curie, due to her great experience in radiochemistry, nearly discovered nuclear fission. In collaboration with the physicist Pavle Savić from Yugoslavia she showed the production of a radioisotope with a half-life of 3.4 h by bombarding uranium with neutrons. In their publication in 1938 they misinterpreted the observation as the discovery of a new element, very similar to lanthanum. On following up these experiments the chemists Otto Hahn and Fritz Straßmann described in 1938 the fission of uranium by neutrons. This finding was based on Lise Meitner's and Otto Frisch's (both physicists) interpretation as a partition of the uranium nucleus into two nuclei with nearly the same mass. A young assistant of Irène remembered Frédéric's remark, that if he had collaborated with his wife, they would have discovered nuclear fission before the German team. After missing the discovery of neutrons, this was the second time Irène Joliot-Curie missed a further Nobel prize by a whisker.

Irène Joliot-Curie was awarded a Nobel Prize in Chemistry in 1935 jointly with her husband Frédéric Joliot-Curie "in recognition of their synthesis of new radioactive elements". Their discovery of artificial radioactivity was not only a great step forward in the development of nuclear physics but also led directly to the possibility of obtaining radioactive isotopes, which are now widely used in medical and biological research and without which nuclear medicine would not exist. Until their discovery, studies of radioactive decay assumed that only naturally occurring substances decompose spontaneously by emission of radioactivity. By bombardment of boron and aluminum with alpha particles, they produced radioactive nitrogen and radioactive phosphorus. These and other analogously produced elements, which are not found in nature, decompose spontaneously within very short periods by emission of positive protons or negative electrons.

When Irène Joliot-Curie obtained her PhD degree in 1925, she was already celebrated as a rising star scientist. Due to her famed name and her mother's support

European Women in Chemistry. Edited by Jan Apotheker and Livia Simon Sarkadi
Copyright © 2011 WILEY-VCH Verlag GmbH & Co. KGaA, Weinheim
ISBN 978-3-527-32956-4

Irène and Frédéric Joliot-Curie working together in their laboratory in the Radium Institut in the late 1920s.

she never had to struggle for career opportunities and succeeded in taking full advantage of them.

Irène got her first job at the Radium Institute, founded by her mother, and stayed there during her entire career. Bensaude-Vincent comments: "Irène never left the niche of the Curie family to venture into an unknown world. It was so obvious to her that she had to follow the same path as her mother, that she never contemplated the possibility of choosing a different way of life."

In the Radium Institute in 1924 she met her husband Frédéric Joliot (1900–1958), one of her mother's PhD students. Close collaboration started after he had finished research for his Doctor of Science degree (1930) and ended five years later with the award of the Nobel Prize in Chemistry.

Their son, like his grandfather Pierre Curie and his father, became a physicist and also was elected to the French Academy of Sciences. His sister Helene Langevin-Joliot was also active in the same field as her parents and grandparents, but like mother and grandmother, she failed to obtain membership of the French Academy of Science.

From her earliest childhood, it was clear that Irène was very intelligent and had exceptional talent in mathematics. To Marie, the education of her daughters was of the utmost importance. When Irène finished her primary school education her mother found no appropriate high school for her and established the 'Cooperative'. In this alternative institution, some of her colleagues, prestigious science scholars, home-schooled their children. For two years Marie Curie taught physics, Paul Langevin mathematics, and Jean Perrin (Nobel Prize in 1926) chemistry. When, after two years, the Cooperative closed, Irène went to Collège Sévigné and passed the baccalauréat just before the outbreak of World War I. During the war Marie Curie managed to set up mobile X-ray machines and helped the medical teams running them. Irène first assisted her mother on the Northern front, later she had her own nursing team.

The first important research project with her husband began in 1931, when they studied the effects of the recent findings of Bothe and Becker who had described a new penetrating radiation, gamma or electromagnetic rays. For that purpose the Joliot-Curies used alpha rays from a very strong Polonium source to bombard thin sheets of different materials. When the materials contained hydrogen they observed a new radiation, which they assumed to be hydrogen nuclei (protons). In consequence of this misinterpretation the Curies missed the discovery of neutrons. James Chadwick (1891–1974) immediately recognized the importance of their results and one month later published some complementary evidence and the discovery of neutrons, for which he got the Nobel Prize in Physics in 1935, the same year when the Joliot-Curies won the Nobel Prize in Chemistry. The distinction between radiochemistry and physics was not very clear at that time. The Radium Institute was very chemistry oriented, in contrast to other European research groups (Cavendish Laboratory (UK), Kaiser Wilhelm Institut (Germany), Fermi's laboratory (Italy), who concentrated on nuclear physics. This may be the reason why other Nobel prizes, for which the Curies provided the basic experimental results, were awarded to members of these laboratories.

After receiving the Nobel Prize, the Joliot-Curie's careers and research collaborations separated. Irène became a professor at the University of Paris and pursued the research program initiated by her parents at the Radium Institute. Frédéric taught at the Collège de France and established his own laboratory. He became the French leader in nuclear physics when he set up a nuclear chain reaction with his collaborators in 1939.

Irène Joliot-Curie's political activities peaked in 1934/35 when she joined the Comité de Vigilance des Intellectuels Antifasciste and, at a time when women had not even obtained suffrage, became "Sous-Secrétair d'État à la Recherche Scientifique" under the Front Populaire, the socialist government of 1936. A feminist motivation to accept the position is assumed, but daily duties were too great for her to meet the demands of the position, which is probably why she resigned after two months. To demonstrate the misogynistic attitude of the French Academy of Science she applied four times for membership between 1951 and 1954, although expecting to be refused. Even her application for membership of the American Chemical Society was rejected in 1953, although for political reasons. As victim of her husband's communistic political activities she also lost her position as *Chef de la Section Chimie* with the CEA, the French atomic energy commission.

Like her mother, Irène suffered from acute leukemia, probably caused by excessive X- and gamma-ray radiation, to which she had been exposed from her early days, during World War I as a radiographer in military hospitals and in the laboratory. Also like her mother, she was not ready to acknowledge the hazards of radioactivity, which were suspected by scientists from the late 1920s on.

Literature

Anonymous (1972) Distinguished nuclear pioneers, Frédéric and Irène Joliot-Curie, *Journal of Nuclear Medicine*, **13** (6), 402–406.

Bensaude-Vincent, B. (1996) Star scientists in a nobelist family. Irène and Frédéric Joliot-Curie. in *Creative Couples in the Sciences*. (eds H. M. Pycior, N. G. Slack, P. G. Abiram) Rutgers University Press, New Brunswick, NJ.

Brain, D. (2005) *The Curies. A Biography of the Most Controversial Family in Science*. John Wiley & Sons, Inc., Hoboken, NJ.

Curie, È. (1952) *Madame Curie*, Fischer Verlag, Frankfurt am Main, Germany.

Jones, L.M. (1990) Intellectual Contributions of Women to Physics. in *Women of Science. Righting the Record*, (eds. G. Kass-Simon and P. Farnes, associate ed. D. Nash) Indiana University Press, Bloomington and Indianapolis.

Maria Kobel (1897–1996)

Annette B. Vogt

■ Maria Kobel was a German chemist and belonged to the first group of women scientists in the Kaiser Wilhelm Society who got the opportunity to become head of a department. Because of the Nazis she had to leave the Kaiser Wilhelm Institute for Biochemistry in 1936. For more than 20 years (until her retirement) she was on the staff of the well-known Beilstein editorial office. Her main research was specifically on tobacco and also more generally on fermentation.

Maria Kobel was born on August 5th, 1897, in Liegnitz, a town in Silesia, the Central European region which is located mostly in present-day Poland. Her family supported her wish to study chemistry, in her memory it became normal for girls to study science in the years after World War I. From 1918 until 1921 she studied chemistry at the University of Breslau (today Wroclaw), and she finished her study with her doctoral thesis (PhD) in 1921 ("Über die in der Literatur als ‚Glyoxylharnstoff' bezeichneten Stoffe" (77 pp.)) under the direction of Johann Heinrich Biltz (1865–1943), a well-known chemist in Germany. Maria Kobel went to Berlin where she was looking for an academic position. She was lucky and was able to obtain a research position in one of the famous Kaiser Wilhelm Institutes (KWI) in Berlin-Dahlem. From 1925 until 1936 she was employed in the KWI for Biochemistry, directed by the founder of this field, Carl Neuberg (1877–1956). At first she held an assistant position, then from 1928 until 1936 she became the head of a department. Her small department "tobacco research" dealt with fermentation research problems. One of the aims was to construct biochemically tobacco tissues with a quality similar to those plants from the Orient, thus helping the German tobacco industry to produce a better quality of tobacco. Although the research topic in Kobel's department was related to applications her research was first related to the basis of fermentation processes. She published several articles in the journal *Biochemische Zeitschrift*, the famous journal which Carl Neuberg founded and edited until his displacement because of the Nazis. Her most important results on fermentation research she published in the volumes edited by Eugen Bamann and Karl Myrbäck in 1941.

European Women in Chemistry. Edited by Jan Apotheker and Livia Simon Sarkadi
Copyright © 2011 WILEY-VCH Verlag GmbH & Co. KGaA, Weinheim
ISBN 978-3-527-32956-4

Maria Kobel (Archive MPG, Berlin: VI. Abt., 1)

Because of the Nazi regime in Germany the situation changed completely in 1933. From 1933 until 1936 several attacks were made by Nazis, against the on the KWI for Biochemistry in general and in particular against the director Carl Neuberg. Finally, he was displaced, and the KWI was closed and re-opened again in 1936 when Adolf Butenandt (1903–1995) became the new director. Maria Kobel, the collaborator and friend of Carl Neuberg, lost her position too (although she was "Aryan"). Thanks to her work and the help of Carl Neuberg she got a position on the staff of the editorial office of the famous series "Beilstein", the handbook, useful for generations of chemists. The Beilstein office was located in Berlin but because of World War II it moved to Frankfurt (Main). Maria Kobel worked on several editions of "Beilstein" until 1962 when she retired. All her life she was remembering the best time she had as a researcher in the KWI for Biochemistry under the direction of Carl Neuberg with whom she was in contact until his death. Maria Kobel died on August 14th, 1996, in Kronberg (Taunus), near Frankfurt (Main) in her 99th year.

Literature

American Philosophical Society, Philadelphia: Nr. 815, Neuberg Papers (letters, 1948–1956). Poggendorff, *Biographisch-Literarisches Handwörterbuch zur Geschichte der exakten (Natur)wissenschaften* Vol. III (1898), IV (1904), V (1926), VI (1936), S. 1346 (Eig. Mitteil.); VIIa (1956), p. 812, VI-Ib (1968ff.), Leipzig u. a.

Archive of the Kaiser-Wilhelm-/Max-Planck-Gesellschaft, Berlin.

Beilstein-Redaktion, Spring 1995 to AV; family Kobel to AV; interviews Maria Kobel with AV, July (5–7.7.) 1995.

Conrads, H. and Lohff, B. (2006) *Carl Neuberg – Biochemie, Politik und Geschichte. Lebenswege und Werk eines fast verdrängten Forschers*, Stuttgart (= Geschichte und Philosphie der Medizin, Vol. 4); Engl. edn, Lohff, B. and Conrads, H. (2007) *From Berlin to New York. Life and Work of the Almost Forgotten German-Jewish Biochemist Carl Neuberg (1877–1956)*, Franz Steiner Verlag, Stuttgart.

Engel, M. (1982) Carl Neuberg, in *Bibliotheks-Information*, Freie Universität Berlin, Vol. 3, pp. 11–16

Engel, M. (1984) Geschichte Dahlems, Berlin

Engel, M. (1994) Paradigmenwechsel und Exodus. Zellbiologie, Zellchemie und Biochemie, in *Exodus von Wissenschaften aus Berlin* (eds W. Fischer et al.) Walter de Gruyter, Berlin, New York, 296–342.

Lieben, F. (1970) *Geschichte der Physiologischen Chemie*, Hildesheim, especially pp. 257, 369, 520.

Maria Kobel's most important publications: 5 articles, together with Eberhard Hackenthal, in: Bamann, E. and Myrbäck K. (Eds) (1941) *Die Methoden der Fermentforschung*, 4 volumes, Leipzig, Vol. 1, pp. 68–73 and 111–115; Vol. 3, pp. 2173–2196, 2197–2205, 2206–2213.

Ogilvie, M., and Harvey, J. (Eds) (2000) The Biographical Dictionary of Women in Science. Pioneering Lives from Ancient Times to the Mid-20th Century, Routledge, New York and London, Vol. 1, p. 711.

Vogt, A. (2008) *Wissenschaftlerinnen in Kaiser-Wilhelm-Instituten. A-Z.*, 2. erw. Aufl., (= Veröffentlichungen aus dem Archiv zur Geschichte der Max-Planck-Gesellschaft, Bd. 12), Berlin, pp. 98–100.

Vogt, A. (2007) *Vom Hintereingang zum Hauptportal? Lise Meitner und ihre Kolleginnen, an der Berliner Universität und in der Kaiser-Wilhelm-Gesellschaft*, Franz Steiner Verlag, Pallas&Athene, Stuttgart, Vol. 17.

Katharine Burr Blodgett (1898–1979)

Sally Horrocks

■ Katherine Burr Blodgett, an American, was the first women to be awarded a PhD in physics from the University of Cambridge. She had earlier been one of the first to work as a scientist in the General Electric (GE) research laboratory at Schenectady, New York, to which she returned after being awarded her PhD. At GE she made significant contributions to industrial chemistry, particularly surface chemistry, where her specialism was thin films. She is credited as the inventor of non-reflecting glass and held a number of patents for her work as well as publishing in academic journals. Her achievements in the field of industrial science were particularly noteworthy among female scientists of her generation, whose efforts to pursue careers in industry frequently met with frustration, but like most industrial scientists she had a limited public profile and her name is not widely recognized outside her field.

Katherine Burr Blodgett was born on 10 January, 1898 in Schenectady, New York the second of two children of George Bedington Blodgett, a GE patent attorney and his wife Katherine Buchanan Burr. Blodgett never knew her father, who died of injuries sustained in a burglary a few weeks before she was born. Most of her childhood was spent in New York City, interspersed with extended stays in both France

Katherine Burr Blodgett.
(Smithsonian Institution Archives).

European Women in Chemistry. Edited by Jan Apotheker and Livia Simon Sarkadi
Copyright © 2011 WILEY-VCH Verlag GmbH & Co. KGaA, Weinheim
ISBN 978-3-527-32956-4

and Germany. After attending Rayson School in New York, which was unusual in that it provided girls with a strong education in mathematics and science, she won a scholarship to Bryn Mawr College, from which she graduated in 1917 with an A.B., with physics as her major subject. A masters degree in chemistry from the University of Chicago followed in 1918.

After completing her studies Blodgett managed to obtain a post as a technician and research assistant at the GE research laboratory in Schnectady, working with Dr Irving Langmuir and earning $125 a month. Although she was not the first female researcher to be appointed by GE it was still unusual for women to secure research positions in industry and the firm's willingness to appoint her may have owed something to the shortage of scientific and technical staff caused by the demands of World War I. In 1924, with the encouragement of Langmuir, she moved to the Cavendish Laboratory at the University of Cambridge as a graduate student attached to Newnham College. Two years later she completed her research and became the first woman to be awarded a PhD in physics by Cambridge. There is evidence that she found the training that she had received from Langmuir essential to her ability to complete her PhD research, and that she did not find the environment at the Cavendish Laboratory to be a supportive one.

After completing her studies Blodgett returned to GE and continued to work with Langmuir, initially on electric lamp filaments. In 1933 the focus of her work moved to thin films, the field in which she did her most important work. This research resulted in numerous academic papers and the majority of the patents filed under her name. This contribution is recognized by the use of the term Langmuir–Blodgett film to refer to layers of organic material one molecule thick, deposited on a solid substrate, the techniques for which she pioneered. After the initial innovation, Blodgett worked on further uses for these films until 1941 when her research was directed towards problems related to national defence. These included using her expertise in surface chemistry to address the problem of icing of aircraft wings, and work on improved smoke screens. She also worked on cloud physics and was a pioneer in the use of computer simulations.

Although she was not a well-known figure among the general public, Blodgett's work received significant recognition from a number of directions, starting in the late 1930s. She was awarded honorary degrees by four institutions, Elmira College (1939), Brown University (1942), Western College (1942), Russell Sage College (1944) and in 1945 won the American Association of University Women Annual Achievement Award. The Garvan Medal of the American Chemical Society followed in 1951, the same year that she was the only scientist to be included in Boston's First Assembly of American Women of Achievement and was chosen by the United States Chamber of Commerce as one of its 15 'women of achievement'. The Photographic Society of America bestowed its Progress Medal on her in 1972 and in 2008 an elementary school in Schenectady was opened bearing her name.

Blodgett, died at her home in Schenetady on 12 October 1979. In her later years she had become involved in the local community and pursued her hobbies of gardening, astronomy and antique collecting.

Scientific Work

Blodgett's initial research at GE with Langmuir was on the properties of incandescent lamps and vacuum tubes. These studies crossed the borders between physics and chemistry and resulted in a number of co-authored publications. Her PhD thesis described the determination of the mean free path of electrons in mercury vapor, a topic that had several applications that would have been of interest to GE. She was known for the careful design of her experiments, the skill with which she carried them out and the thoroughness with which she analyzed her results.

These attributes were crucial to the success of her work on thin films during which she pioneered a technique for transferring a series of monomolecular films of oil from the surface of water to a solid surface. Her discovery meant that the thickness of these films could be controlled with great accuracy and opened up a new field of research that offered the possibility of many practical applications that she and others worked to exploit over subsequent decades. This research on thin films was first reported in a brief communication to the *Journal of the American Chemical Society* in 1934 and expounded in detail in a longer paper, 'Films built by depositing successive monomolecular layers on a solid surface' published in the same journal the following year. Here, she described how she had found it possible to deposit more than 200 layers on glass and various metals and described experiments in which the thickness of these films was measured by the interference of monochromatic light reflected by them. This observation led to the development of the Blodgett Color Gauge. This enabled researchers to measure the thickness of a film only a few millionths of an inch thick by comparing the colors in the gauge with the color of their film. GE marketed this for use in laboratories in place of more sophisticated and expensive optical apparatus.

The nature of these films meant that they could be controlled to within a fraction of the typical wavelength of visible light and suggested that it might be possible to exploit this to produce an 'invisible' glass, the second and most important application of this research. This could be achieved if light reflected from the underlying glass layer could be cancelled out by light reflected from the top of the film and this basic principle was reported in *Physical Review* in 1939 in a paper entitled 'Use of Interference to Extinguish Reflection of Light from Glass'. This work provided a foundation for the development of antireflective coatings used in situations such as cameras and telescopes where reflection impaired performance. Antireflective glass was also adopted in art galleries for the exhibition of paintings and photographs. The commercial exploitation of this glass was made possible once more permanent films than those initially used by Blodgett in her research had been developed by others.

Blodgett's wartime research was followed by further collaborations with the US military in the later 1940s, notably the development of an instrument that could be carried aloft by weather balloons to measure humidity in the upper atmosphere for the Army Signal Corps. Her use of computer simulations to study the trajectory of fine particles in the vicinity of fibers was an early example of the application of computers to research problems and contributed to the development of an efficient aerosol filter. Her final published work was a joint publication with Dr T. A. Vanderslice on the clean-up of gases in ionization gauges.

Katharine Burr Blodgett enjoyed a long and productive career as an industrial scientist at a time when very few women were successful in this field. Unlike the majority of industrial scientists her work was recognized far beyond the firm for which she worked. This visability stemmed primarily from her work on Langmuir–Blodgett films that opened up a major field of study and led to numerous practical applications. Her achievements were closely associated with Irving Langmuir, her mentor, whose faith in her experimental abilities was important in giving her opportunities that few women enjoyed. These included not just her successful career in industrial science but her acceptance to study at the Cavendish Laboratory at the University of Cambridge, where she was the first woman to be awarded a PhD.

Literature

Blodgett, K.B. (1935) Films built by depositing successive monomolecular layers on a solid surface, *Journal of the American Chemical Society*, **57** (6), 1007–1022.

Blodgett, K.B.(1939) Use of interference to extinguish reflection of light from glass. *Physical Review* **55** (4), 391–404.

Gaines, G. Jr. (1980) In memoriam: Katherine Burr Blodgett, 1898–1979 *Thin Solid Films*, **68** (1), vii–viii.

Wise, G. (1999) Katharine Burr Blodgett, in *American National Biography*, vol. 3, Oxford University Press, New York.

Antonia Elizabeth (Toos) Korvezee (1899–1978)

Marianne Offereins

■ The first female Professor at the Polytechnic (now University) of Delft.

On March 8, 1899 in the Frisian village Wijnaldum, a second daughter was born to the Dutch Reformed minister Willem Korvezee and his wife Baukje Andringa. She was christened Antoinia Elizabeth. A year later, the family left Friesland, because the father Korvezee's ideas were too socialistic for the community. The family settled in The Hague. Here the girls grew up and after their primary school they entered the Third Municipal HBS. This school was known as a 'boys' school. Here, Toos attracted attention because of her high marks in mathematics. In 1917 she opted for training as a chemical engineer at the Technical University in Delft. She became a member of the Delft Female Students Association.

Toos Korvezee (Gemeentelijke Archiefdienst Delft).

European Women in Chemistry. Edited by Jan Apotheker and Livia Simon Sarkadi
Copyright © 2011 WILEY-VCH Verlag GmbH & Co. KGaA, Weinheim
ISBN 978-3-527-32956-4

In January, 1922, Toos Korvezee earned her bachelor's degree. Almost immediately after, before the summer of that year, she graduated cum laude in inorganic chemistry under the tuition of Professor F.E.C. Scheffer. For the next two years she had a temporary appointment as an assistant in analytical chemistry in Delft, after that she worked from 1924 to 1938 with Professor Scheffer. During this period she also wrote her doctoral dissertation. On June 5, 1930 Antonia Elizabeth Korvezee attained her doctoral degree with a thesis entitled: *'Koperchloride als katalysator voor het Deacon-proces'* (Copper chloride as a catalyst for the Deacon process). Besides that, she had a large number publications: before 1940 she had written more than 40 papers. Scheffer initially published them as co-author, later publications were often in association with one of the – mostly female – assistants. Most of these articles appeared in the *Recueil des Travaux Chimiques des Pays-Bas*. Toos Korvezee wrote in a brief and businesslike style and often focused on precise measurements.

In the period between 1930 and 1932 Toos twice spent twice a half year in Paris, where she did research on radioactivity in the laboratory of Marie Curie.

In 1935, the temporary appointment of Toos Korvezee was converted into a permanent one. In addition, she was admitted as a private tutor in radioactivity at the Technische Hogeschool (HN). This meant that she worked *'buiten bezwaar van de schatkist'* (*with no charge for the treasury*), that is, she was not paid. Often such a private position led to a position as a professor. Her inaugural lecture was entitled: "The determination of the content of radium compounds".

In 1936, Toos Korvezee was nominated as the first candidate for the vacant chair in analytical chemistry in the chemical engineering department. The nomination was by five votes for her to four against. Although in the nomination 'her very special talent and wit' and 'her clear argumentation and her broad interests' were praised, she was not appointed. About a year later, however, she became principal assistant to Professor Scheffer. In 1940 another chair became vacant. For the chair of Professor of Physical Chemistry Toos stood in third place. She was not appointed, but the place was given to Professor W.G. Burgers.

During the war, work at the Technical University became increasingly difficult. In September 1943 Toos moved to Venlo on unpaid leave, in order to work in a research position with the 'Pope' lamp factory. Later she would explain this departure as a form of protest against the actions of the German occupiers.

After the war, Toos returned to Delft, now as a curator. In September 1948 Toos was appointed lecturer in theoretical chemistry, in this she was the second female lecturer at the TH, after Mrs. J.H. van Leeuwen. Although she specialized in radioactivity, she was not really involved in the preparations for the later reactor center. Toos was especially active in the field of education, among others she taught static thermodynamics.

In 1953, her mentor Scheffers retired, and, again, someone else became professor. A man who was ten years younger, because the function was considered 'a too heavy task' for Toos, although she was seen as 'entirely competent'.

Some time later she received the offer of an associate professorship in theoretical chemistry 'in a personal capacity' and on the same salary as a lecturer. Her own laboratory, however, did not belong to the function. Therefore, Toos moved to the laboratory of Professor W.G. Burgers who in 1940 had been appointed Professor in preference to her.

Toos Korvezee, was appointed as the first female professor at the Polytechnic of Delft in 1954 and gave her inaugural speech on April 14, 1954 her with the subject 'Life and works of Dr. J.J. van Laar'. The rector greeted her arrival by saying: "At the more than one hundred years old aloe, finally a flower blooms".

That this final appointment had not come without a struggle was evident from the fact that even years after the discussion was not silenced. During her professorship Toos supervised five PhD students and more than thirty alumni, despite the fact that she did not have her own staff and no private laboratory. Of course her nomination drew the attention of the press, and when, in 1955, Queen Juliana visited the TH, Professor Korvezee headed the cortege (parade of professors in robes).

In 1989, the Delft Technical University established an annual emancipation prize, named after Antonia Korvezee.

Literature

Bosch, M. (1994) *Het geslacht van de wetenschap. Vrouwen en hoger onderwijs in Nederland 1878–1948*, SUA, Amsterdam.

Damme-van Weele, M. and Ressing-Wolfert, J. (1995) *Vrouwen in techniek. 90 jaar Delftse vrouwelijke ingenieurs*, Delftech, Delft.

Hart, Joke 't (1986) *Een barst in het bolwerk. Vrouwen, natuurwetenschappen en techniek*, SUA, Amsterdam.

Jong, F. de (2002) Biografie van Korvezee, Antonia Elisabeth. *Biografisch Woordenboek van Nederland* 5, Instituut voor Nederlandse Geschiedenis, Den Haag.

Jong, F. de (1988) "Die aloude aloë toch…" A. E. Korvezee (1899–1978), de eerste vrouwelijke hoogleraar aan de Technische Hogeschool Delft. in *Geleerde Vrouwen, Negende Jaarboek Vrouwengeschiedenis*, Seventy years of women's studies (SUN), Nijmegen.

Kolf, dr. Marie van der (1950) *Zeventig jaar vrouwenstudie*, Rotterdam.

Schenk, dra. M.G. (1948) Vrouwen van Nederland 1898–1948. De vrouw tijdens de regering van koningin Wilhelmina, Scheltens & Giltay, Amsterdam.

Mária de Telkes (1900–1995)

Éva Vámos

■ The physicist Mária de Telkes was well known in the USA, where she lived from the age of 24 for over 70 years, while she remained completely undiscovered in her native country, Hungary. She was one of the pioneers of the utilization of solar energy for heating, whereby she invented a process for preserving this energy by utilizing the heat of solidification of the generally known and cheap compound sodium sulfate decahydrate (known also under the name of Glauber salt, and also used as a laxative).

She patented a process for a cooking stove heated by solar energy as well as a method for desalinating sea-water using solar power. Lives of torpedoed sailors and pilots of aircraft shot down and landing in the sea could be saved by this process during World War II. In peace-time the process could be used to ensure the water supply of poor or arid regions.

Mária de Telkes was born in Budapest as the eldest of a well-to-do bank manager's eight children. After primary and secondary school studies she enrolled in the Faculty of Philosophy of Budapest University of Sciences. She studied mathematics and physics, and after graduation became an assistant to the professor of physics István Rybár. She obtained her doctoral degree in physical chemistry in 1924. In the same year her uncle Ernö Ludwig, who happened to be Hungarian consul in Cleveland and was married to an American lady, invited her to the United States. She accepted the invitation, moved to America, and stayed there for over 70 years, that is, for her lifetime. She returned to Hungary only to die in her native country.

She started her career at the research laboratory of the Cleveland Biophysical Institute under the leadership of Professor G.W. Crile. At the Institute the radiation of brain cells was investigated. They were able to measure the infrared radiation of the brain cells with an electric camera devised by Mária Telkes. They also investigated the potential difference in animal tissues and found, among other things, that the life of an organism existed only as long as potential differences were maintained within the organism.

European Women in Chemistry. Edited by Jan Apotheker and Livia Simon Sarkadi
Copyright © 2011 WILEY-VCH Verlag GmbH & Co. KGaA, Weinheim
ISBN 978-3-527-32956-4

Mária de Telkes. This image is available from the United States Library of Congress's Prints and Photographs Division under the digital ID cph.3c13268.

In 1934 The New York Times published a list of the 11 best known women of the US: besides film stars, sportswomen and other personages of public life Mária was the only scientist mentioned.

In 1939 she moved to Boston, where she started her real career, focused around the utilization of solar energy, as researcher and professor at MIT (Massachusetts Institute of Technology). A project financed by private capital was started in 1938. The project was aimed, in the first place, at utilizing/transforming solar energy. The project funding of $650 000 served to cover research and implementation of the results until 1988, and was financed by the tycoon Godfrey Lowell Cabot. Mária was invited to participate in the project, of which she became the leader in 1940.

Using solar energy for heating instead of fossil fuels would mean considerable savings. The greatest problem in utilizing solar energy for heating was the storage of the sun's heat. In order to overcome this difficulty, Mária Telkes used a process of chemical heat storage. She found that the best compound for the purpose was a quite common and cheap chemical, sodium sulfate decahydrate, commonly known as Glauber salt. The solubility of the compound increased more than tenfold between 0 and 32.4 °C, and remained nearly stable with increasing temperatures up to 200 °C, whereby its melting heat was quite high at 32.4 °C. The molten compound could preserve the absorbed solar energy for up to 10 days, and release it upon cooling. The phase transition heat released upon solidification of sodium sulfate was 82 times higher than that of water.

Six "sun houses" were built under the project. The first house devised for solar heating in 1948 utilized the principle of heat preservation described above. The building itself was designed by the architect Eleanor Raymond and the heating system by Mária Telkes. Her relatives, the Némethy family, were the inhabitants of the experimental house. Tanks containing the salt solution were mounted on top of and on the sides of the house, while the heating of the inside was provided by a system of pipes in which air and water circulated. However, the system failed in very cold winter time, as was experienced as soon as the winter of 1948, when the heat generated by the system was insufficient to ensure a desirable temperature within the house. With regard to the climate of the northern parts of the US an auxiliary (traditional) heating system should have been built for such cases. This solar heating system was finally dismantled in 1953 but others followed and made Mária de Telkes's name well known all over the US.

Mária de Telkes published over 100 papers, and over 20 patents.

One of her most important patents dealt with the desalination of sea water by solar energy, a process mainly used in tropical regions.

She solved the problem of the storage of the cold by the same principle as for heat storage. In this way houses could be air conditioned by solar energy. A patent of hers, granted when she was 90 years old, also dealt with the problem of storing the cold.

Her solar-energy-based cooking oven became widespread in India as it was easy to handle and inexpensive, and there was sufficient sunshine.

Besides being a university professor and a successful inventor Mária was also a consultant to a number of industrial companies and even participated in space research.

She obtained 12 distinctions for her work. The first, in 1927, was however awarded to her for life-saving: while spending her holidays on the shores of Lake Erie, she noticed a wooden house on fire and a woman running out of it crying as her little daughter remained in the house. Risking her life, Mária ran into the blazing house and saved the child.

Another distinction worth mentioning was awarded to her by the Society of Women Engineers, Washington. The reasons for awarding the distinction were given by the society as follows: "Society of Women Engineers' Award to Maria Telkes in recognition of her meritorious contribution to the utilization of solar energy." It seems, perhaps, strange, that in her homeland she was mentioned only in short communications in newspapers, and before World War II.

She had no family of her own in the USA. Apparently, and contrary to many emigrants, she was never tortured by homesickness. She returned to Hungary but once, at the age of 95, in 1995, and died in the same year, on December 2. Apparently this was not considered by her country as very important news as it reached the US with considerable delay. That is why a detailed obituary was only published in the Rocky Mountain News on August 19, 1996, nearly nine months later.

Mária Telkes was a very talented, versatile and inventive scientist. She also had the gift of recognizing the utility of her ideas and the ways that they could be put into practice. Her versatility is proven by the fact that when working in a team of biophysicist she promoted their work by giving them a tool – an electric camera – of her own design to enable them to reach the goal they had set themselves. When she moved, after working for over a decade in one field, to another city, she was able to simultaneously switch to an entirely new field of research. The utilization of solar energy has engaged the attention of people throughout the history of mankind. However, a solution that would allow the economical use of this inexhaustible source of energy had not been found earlier. Mária dealt with the problem as a real scientist. Instead of entering the path of trial and error she elucidated the theoretical roots of the problem, finding a solution (namely the use of sodium sulfate) that might seem very simple, nevertheless it had not occurred to anyone earlier. This one idea formed the basis of practically all her inventions, from sun-houses through desalination of sea water to cooking ovens.

When looking at her life's work we cannot but admire the wealth of her ideas and her determination to put them into practice for the benefit of people, to make their lives easier and more comfortable, and sometimes even to save them. She was, no doubt, one of the most significant and successful scientists of the twentieth century and not only of the female ones. If we consider that today we are permanently searching for new and renewable sources of energy, that will not be depleted in the foreseeable future, we must admit that her research was much ahead of her time. It is, perhaps, of greater value to people of the twentyfirst century than it was to those of the twentieth.

Literature

Crile, G.W., Rowland, A.F. and Telkes, M. (1928) An interpretation of excitation, exhaustion and death in terms of physical constants. *Proceedings of the National Academy of Sciences*, 532–538.

Cattell, J. (ed.) *American Men & Women of Science. Physical and Biological Sciences*. 16th edn, vol. VII. T–Z, R. R. Bowker Co., New York, p. 57.

Fields, S. (1964) Harnessing the Sun. *Daily News Florida*, June 19, 1964.

Pap, J. (1997) A napenergia magyar tudósnöje, a szolártechnika nagyja, dr. Telkes Mária. (Hungarian scholar of solar energy, great personage of solar technology, Dr. Maria Telkes) in *Tanulmányok a Természettudományok, a Technika és az Orvoslás Történetéből* (Studies in Sciences, Technology and Medicine). MTESZ and OMM, Budapest, pp. 43–45.

Pap, J. (2000) A napenergia magyar tudósnöje, a szolártechnika nagyja, dr. Telkes Mária. (Hungarian scholar of solar energy, great personage of solar technology, Maria Telkes), in *Asszonysorsok a 20. Században*. (Fates of women in the 20th century). (eds M. Balogh, and K. Nagy) BME Szociológia és Kommunikáció Tanszék, Szociális és Családvédelmi Minisztérium Nöképviseleti Titkársága, Budapest, pp. 79–83.

Rédey, S. Telkes Mária – Az ismeretlen Napkirálynö. (Maria Telkes – The Unknown Sun Queen), Természet Világa, Budapest. http://www.termeszetvilaga.hu/szamok/tv2009/tv0903/redey.html. 2010/07/07, 5 p.

Saxon, W. (1996) Maria Telkes, 95, An Innovator of Varied Uses for Solar Power. *New York Times*, August 13, 1996.

Saxon, W. (1996) Maria Telkes, Hungarian–American Solar-Energy Advocate and Pioneer. *Rocky Mountain News*, August 19, 1996.

Society of Women (1952) *Engineers, Award Committee: Date of Award: March 15, 1952*; S. W. E. Convention, New York, NY.

Ujfaludi, L. (2003) A napenergia-hasznosítás roved története. (Short history of solar energy utilization.) *Fizikai Szemle*. (Review of Physics), **3**, 99–114. http://www.kfki.hu/fszemle/archivum/fsz0303/ujfal0303.html.

Erika Cremer (1900–1996)

Annette B. Vogt

■ Erika Cremer was a German physical chemist and Professor Emeritus at the University of Innsbruck and Corresponding Member of the Austrian Academy of Science (in 1964). She was one of the most important pioneers in gas chromotography. She first conceived the technique in 1944.

Erika Cremer was born in Munich, on May 5th, 1900. She grew up in the family of the physician and university professor Max Cremer (1865–1935) who supported her very much and encouraged her not only to study but to obtain a doctoral degree. Her elder brother Hubert Cremer (1897–1983) became a mathematician and university professor (1949–1966) at the Technical College Aachen.

Erika Cremer (in: Wöllauer (1997), Oberkofler (1998),
Beneke (1999)
Deutsches Museum Bonn, Katalog 1995, S. 308 (Photo),
S. 311 Skizze des Chromatographen)

European Women in Chemistry. Edited by Jan Apotheker and Livia Simon Sarkadi
Copyright © 2011 WILEY-VCH Verlag GmbH & Co. KGaA, Weinheim
ISBN 978-3-527-32956-4

After high school education she studied chemistry from 1921 until 1927 at the Berlin University, the Friedrich-Wilhelms-Universität, and received her doctoral degree in 1927 with the thesis "Über die Reaktion zwischen Chlor, Wasserstoff und Sauerstoff im Licht". Her supervisor for her doctorate studies was the famous chemist Max Bodenstein (1871–1942). She wanted to become a scientist, and although it was still difficult for women to do so she was able to become a scientist and carry out research all her life. From 1927 until 1940 she worked in different research institutions, partly with grants. From 1928 until 1930 she obtained a fellowship to work with György (Georg) Hevesy (1885–1966) at the University Freiburg i. B., from 1930 until 1934 she did research in Berlin, from 1934 until 1937 she earned a fellowship to work with Kasimir Fajans (1887–1975) at the University of Munich, and from 1937 until 1940 she did research again in Berlin at the Kaiser Wilhelm Society. She learned that as a woman she had to overcome more difficulties than her male colleagues, but thanks to the support of her family, as well as of some professors, she worked successfully. It was only in 1940 that she got her first academic position at the University of Innsbruck (at that time occupied by the Nazis).

Erika Cremer belonged to a small group of women scientists who were employed at different times in different Institutes of the Kaiser Wilhelm Society (established in 1910 as the first research institution in Germany, and which soon became one of the most prestigious institutions), namely in three Kaiser Wilhelm Institutes (KWI). In the winter of 1927/28 and from 1930 until 1933 she held unofficial assistant positions in the KWI for Physical Chemistry and Electrochemistry, directed by Fritz Haber (1868–1934) in Berlin-Dahlem. She belonged to the department of Michael Polanyi (1891–1976) and investigated para- and ortho-hydrogen in the special laboratory (Kältelaboratorium) of the Physikalisch-Technische Reichsanstalt. Because of the Nazi regime in Germany many scientists – both female and male – were displaced (as the Academic Assistance Council in London named it) from institutions, including from the Kaiser Wilhelm Institutes. Michael Polanyi was one of those dismissed, his department was closed and Erika Cremer lost the possibility for further research there. After her work in Munich, from April 1937 until December 1937, she became a "private assistant" of Otto Hahn (1878–1968), director of the KWI for Chemistry in Berlin-Dahlem and head of the chemical department. Otto Hahn was a friend of Kasimir Fajans as well as a close friend and coworker of the famous physicist Lise Meitner (1878–1968) who was the head of the physical-radioactivity department in the same KWI for Chemistry until her escape into exile in July 1938. Finally, from 1939 until 1940 Erika Cremer held an assistant position in the KWI for Physics, in the department of Karl Wirtz (1910–1994). During her stay in Berlin she continued her research on para- and ortho-hydrogen and wrote her Habilitationsschrift to become a Dozent at the Berlin University. The procedure Habilitation was finished successfully on February 10th, 1939. The title of her paper (Habilitationsschrift) was "Bestimmung der Selbstdiffusion in festem Wasserstoff aus dem Reaktionsverlauf der Ortho-Para-Umwandlung". She was not nominated as a Dozent at the Berlin University but she became a Dozent at the University of Innsbruck in 1940.

At the University of Innsbruck she started her research on gas chromatography. Between 1944 and 1947 she developed a special method to separate gases by utilizing an inert carrier gas. She undertook this research together with her doctoral student Fritz Prior (1921–1996), who received his doctoral degree in 1947. Later he became a politician in Austria. After the liberation of Austria and the capitulation of Nazi Germany in May 1945, Erika Cremer was able to stay at the University of Innsbruck. In 1945 she became the head of the physical-chemical Institute, in 1951 Professor and in 1959 Full Professor. From 1953 until 1954 she was a guest scientist at MIT in Cambridge MA, USA. Erika Cremer died in Innsbruck on September 21st 1996.

Erika Cremer was one of the first female chemists who made a successful academic career in the 20th century, and she was highly acknowledged internationally. She received several honours. In 1964 she was elected as Corresponding Member of the Austrian Academy of Science. In 1958 she was awarded the Wilhelm Exner Medal, in 1970 the Erwin Schrödinger Prize, and in 1977 and 1978 the Tswett Medal of the USA and in the USSR. She became one of the most famous women scientists in Austria. In 1995 the Deutsches Museum mounted an exhibition in its branch in Bonn, explaining to the public how Erika Cremer built the first gas chromatograph with Fritz Prior in the 1940s.

Literature

Archive of the Austrian Academy of Science, Vienna.

Archive of the Kaiser-Wilhelm-/Max-Planck-Gesellschaft, Berlin.

Archive of the University of Berlin, (files on the thesis and about the Habilitation).

Archive of the University of Innsbruck.

Beneke, K. (1999) Erika Cremer, in *Biographien und Wissenschaftliche Lebensläufe von Kolloidwissenschaftlern, deren Lebensdaten mit 1996 in Verbindung stehen*, Reinhard Knof Verlag, Nehmten, pp. 311–334 (list of publications by Erika Cremer pp. 330–334).

Deutsches Museum Bonn (1995), Katalog der Ausstellung 1995, pp. 308–311.

Keintzel, B. and Korotin, I. (Eds) (2002) *Wissenschaftlerinnen in und aus Österreich. Leben – Werk – Wirken*, Böhlau, Wien, pp. 121–124.

Miller, J.A. (1993) E. Cremer in *Women in Chemistry and Physics. A Biobibliographic Sourcebook* (eds L. S. Grinstein, R. K. Rose, and M. H. Rafailovich), Greenwood Press, Westport, Connecticut, London, pp. 128–135.

Oberkofler, G. (1998) *Erika Cremer: Ein Leben für die Chemie*, Studien-Verlag, Innsbruck-Wien.

Ogilvie, M. and Harvey, J. (Eds) (2000) *The Biographical Dictionary of Women in Science. Pioneering Lives from Ancient Times to the Mid-20th Century*, Vol. 1, Routledge, New York and London, p. 301–302.

Video (documentary film with Erika Cremer, produced in 1989/1990, 45 minutes).

Vogt, A. (2008) *Wissenschaftlerinnen in Kaiser-Wilhelm-Instituten. A-Z*, 2. erw. Aufl., Berlin, (= Veröffentlichungen aus dem Archiv zur Geschichte der Max-Planck-Gesellschaft, Bd. 12), pp. 44–46.

Vogt, A. (2007) *Vom Hintereingang zum Hauptportal? Lise Meitner und ihre Kolleginnen an der Berliner Universität und in der Kaiser-Wilhelm-Gesellschaft*, Franz Steiner Verlag, Pallas&Athene, Vol. 17.

Wöllauer, P. (1997) Wir müssen leider eine Frau nehmen, ... Erika Cremer und die Entwicklung der Gaschromatographie, *Kultur Technik* 1, 29–33.

Elisa Ghigi (1902–1987)

Marco Ciardi and Miriam Focaccia

■ Elisa Ghigi, Professor in organic chemistry at the University of Bologna, had a firm long-term commitment to experimental research work and spent the majority of her life in the laboratory. The results are documented by a huge scientific output (more than 100 publications) in different areas of organic chemistry. Most of these appeared in the *Gazzetta Chimica*, a journal that enjoyed international prestige despite being published in Italian, although some were also printed in the *Berichte der Deutschen Chemischen Gesellschaft* (1937, 1938) and in *Helvetica Chimica Acta* (1940). She mainly dealt with natural organic substances and, as far as organic synthesis is concerned, she focused on nitrogen heterocyclic compounds (pyrroles, indoles and carbazoles) by preparing new compounds, as well as acenaphthene derivatives.

Elisa Ghigi was born in Bologna on 25 June, 1902 into a wealthy family of chemists. The Casati Law in 1859 introduced a physics–mathematics course in technical colleges, which was recognized for enrolling in university faculties of mathematics, physics and natural sciences and schools of pharmacy. The law did not exclude women from access to the higher reaches of education, because the idea had not even been taken into consideration; this left the way clear for the first female students to lay claim to the right to study. Elisa Ghigi was one of them and was awarded a physics–mathematics diploma in 1920.

In addition to the debates on women's rights that were raging at the time, a large number of women also started going to university from the 1870s onwards, including those who studied pharmacy, thanks to the new Bonghi Law that governed the attendance of female students at university.

The University School of Pharmacy was opened in Bologna, in 1859 and offered two courses: a diploma in Pharmacy that lasted three years plus one year of pharmaceutical experience to be spent in a chemist's, and a degree in Chemistry and Pharmacy, awarded at the end of four years of theoretical studies and one year of practical work experience. Elisa graduated in Chemistry and Pharmacy on 6 July 1925, with top marks "cum laude". Her thesis, entitled "Phenylpyrroles and some of their azoic and aldehydic derivatives", was carried out under the tutelage of Professor Giuseppe Plancher, himself a student of Giacomo Ciamician and an illus-

European Women in Chemistry. Edited by Jan Apotheker and Livia Simon Sarkadi
Copyright © 2011 WILEY-VCH Verlag GmbH & Co. KGaA, Weinheim
ISBN 978-3-527-32956-4

Elisa Ghigi (picture received from authors).

trious expert in organic synthesis. Her other tutors were Gaetano Charrier (whose obituary she later wrote in 1938) and Giovan Battista Bonino; thanks to them Elisa built up a solid scientific background.

In 1933 she qualified as a university teacher in pharmaceutical chemistry and taught many courses over the years, including organic chemistry, in faculties of pharmacy and science.

In 1948 she was awarded a professorship in Pharmaceutical Technology at the Faculty of Pharmacy at the University of Ferrara, while the following year she started teaching organic chemistry at the same university and also became Head of the Institute.

In 1952 she was awarded the professorship in organic chemistry that her tutor, Gaetano Charrier, had held and stayed at the University of Bologna for the rest of her academic career.

From 1956 to 1965 she was Head of the Faculty of Pharmacy. It was thanks to her dedication and her perseverance in the search for funds that it was possible to renovate the Institute of Pharmaceutical Chemistry and equip it with modern laboratories and a well-stocked library.

She published her early works regarding pyrrole and its derivatives with Charrier, but soon ventured into studies on carbazole derivatives by herself. In 1932 she started a series of projects with Charrier. Among other things they focused on the formation of benzanthrone and its derivatives. In around 1938 her work took on a more specifically chemical-pharmaceutical nature and she started publishing in the *Annali di Chimica Farmaceutica*. She focused on Vitamin B6, procaine and digital technology. Elisa Ghigi ventured into the difficult issue of the formation of Marrubiin, a substance contained in the leaves and shoots of Marrubium Vulgare (Horehound), a labiate plant which is especially common in the Salento peninsula. She published some 11 papers on the subject, the first with Bernardi in 1947 and the last in 1956. Her research was based on a long complex sequence of reactions and separations, which took a long time and required much effort. It was difficult to draw the conclusions now reached with much less effort by modern means. Towards the end of her career she focused on the structure and synthesis of simonellite, a terpenoid fossil which is still studied by geochemists, as well as acenaphthene and its derivatives, also by using infrared spectroscopy.

Although she was an attentive passionate teacher who enjoyed excellent relations with her students, who appreciated her kindness and goodness, she seemed indecisive in embracing the major innovations that revolutionized organic chemistry in the teaching field in the middle of the century. She published *Organic Chemistry Lectures* (Bologna, 1953) and a year later *Treatise of Inorganic Pharmaceutical Chemistry* (Bologna, 1954) with Charrier. Unfortunately, rapid progress in the field of organic chemistry and the advent of reaction mechanisms soon made her "Lectures" obsolete. They are, however, still useful for obtaining information on the origin of certain natural compounds.

Literature

Archivio Storico dell'Università di Bologna (www.archiviostorico.unibo.it).

Colonna, M. (1990) Elisa Ghigi e la sua opera, in *1737–1987. Dalla Cattedra di J. B. Beccari ai Dipartimenti. 250 Anni di Chimica*, (eds Breccia Fratadocchi, A. and Pasquinelli, A.) Università di Bologna, Bologna.

Dalla Casa, B. and Tarozzi, F. (1988) Da "studentinnen" a "dottoresse": la difficile conquista dell'istruzione universitaria tra '800 e '900, in *Alma Mater Studiorum, La Presenza Femminile dal XVIII al XX secolo. Ricerche sul Rapporto Donne/cultura Universitaria Nell'ateneo Bolognese*, CLUEB, Bologna, pp. 164–165.

Raicich, M. (1989) Liceo, università, professioni: un percorso difficile, in *L'educazione Delle Donne: Scuole e Modelli di Vita Femminile Nell'Italia Dell'Ottocento*, (ed. S. Soldani) Franco Angeli, Milano, pp. 147–181.

Kathleen Lonsdale (née Yardley) (1903–1971)

Sally Horrocks

■ Kathleen Londsdale was initially trained in mathematics and physics before moving into chemistry through her research in X-ray crystallography, then a new field which, in Britain, attracted and retained an unusually high number of very talented women scientists. Lonsdale's mentor was W.H. Bragg, who secured her research funding after she graduated and who continued to support her career when marriage and motherhood seemed to threaten her ability to continue with scientific work. The other key man in her life was her husband, Thomas Lonsdale, whose own willingness to take on domestic tasks, then seen as exclusively women's work, was also crucial. It not surprising that Lonsdale, in later life, advised women seeking careers in science to choose their husband with care. Had she not done so it is unlikely that she would have managed more than a fraction of what she achieved. Thomas was also her partner in her other career as a prison reformer and peace activist.

Kathleen Yardley was born in Newbridge, Ireland on 28 January, 1903. She was the youngest of the ten children of Harry Frederick Yardley, who was the local postmaster, and his wife, Jessie Cameron, a strong character despite her small stature. When Kathleen was five her mother, originally from London, decided that the family should leave Ireland for the more stable environment to be found in Essex. After attending Downshall Elementary Schools between 1908 and 1914 Kathleen won a scholarship to the Ilford County High School for Girls, where she remained until 1919, also taking classes in physics, chemistry and higher mathematics at the County High School for Boys. This made it possible for her to secure a scholarship to study at Bedford College for Women in London when she was just 16. Here she initially studied mathematics before changing to physics because of her enthusiasm for laboratory work and the prospects this seemed to offer of a career in experimental research rather than teaching, the occupation most easily accessible to women science graduates during the interwar years. In 1922 she scored the highest mark in the University of London BSc examinations. This achievement brought her to the attention of W.H. Bragg, one of the examiners, who offered her a research position in X-ray crystallography at University College, London where she

European Women in Chemistry. Edited by Jan Apotheker and Livia Simon Sarkadi
Copyright © 2011 WILEY-VCH Verlag GmbH & Co. KGaA, Weinheim
ISBN 978-3-527-32956-4

Kathleen Lonsdale (http://www.britannica.com/EBchecked/
topic-art/347705/39025/Dame-Kathleen-Lonsdale-1948).

received a Department of Scientific and Industrial Research grant of £180 a year. When Bragg moved to the Royal Institution (RI) in 1923, his research team including Yardley, moved with him. The following year she published her first papers.

In 1927 Yardley married Thomas Jackson Lonsdale, an engineer who she had met during her time at University College. They moved to Leeds where Thomas had a post as an assistant at the Silk Research Association, housed in the Textile Department of the University of Leeds. Although it was usual for women to retire from scientific work on marriage, Kathleen continued her research with the support of her husband, who is said to have taken the view that he had not married to get a free housekeeper. Between them they ensured the domestic tasks were both minimized and shared, so that they were both able to continue with their scientific work at home in the evenings. Kathleen worked in the Department of Physics where she had a part-time demonstrator's post to supplement the Amy Lady Tate scholarship that she had been awarded by Bedford College for 1927–1929. A grant from the Royal Society enabled her to buy new equipment and Professor C.K Ingold of the Leeds Department of Chemistry provided crystals of hexamethylbenzene, whose structure she worked to elucidate.

The Lonsdale's first child, Jane, was born in 1929 and shortly afterwards they returned to London where Thomas was employed at the Road Research Laboratory. Kathleen continued with her research throughout her pregnancy. When her daughter was small a grant of £50 from the Managers of the Royal Institution enabled her to employ a daily help, giving her time to work on her calculations. The return to London had, however, disrupted her experimental work and for the next two years she continued to work at home on more theoretical and mathematical problems. Her second child, Nancy, was born in 1931 and in the same year Bragg managed to secure sufficient funding from Sir Robert Mond to pay Lonsdale enough money to entice her back into his laboratory, where she stayed for the next 15 years. Initially she was Bragg's research assistant but later secured grants and fellowships for her work. Her third child, Stephen, was born in 1934. In 1936 she was awarded a DSc by the University of London. When Bragg died in 1942 she continued to work at the RI under Sir Henry Dale and was a Dewar Fellow from 1944 to 1946. Dale, as President of the Royal Society, was instrumental in smoothing the way for women to be elected to the Fellowship, and in 1944 Lonsdale, along with Marjorie Stephenson, was nominated. On 22 March, 1945 these two were the first women to be elected to the Fellowship of the Royal Society.

In 1946 Lonsdale made the transition to a university department when she accepted the post of Reader in Crystallography at University College, London. She had previously withdrawn from the selection process for the Chair of Physics at Bedford College when she realized the amount of teaching that would be required. She also became editor of the *International Tables for X-ray Crystallography*, a project that absorbed considerable amounts of her time, perhaps detracting from her ability to pursue her own research. In 1949 she was promoted to Professor and Head of Department, enabling her to develop her own research school. She established a successful course in crystallography for chemistry undergraduates and an inter-collegiate MSc course in collaboration with J.D. Bernal at Birkbeck College. She continued with her existing lines of research and later developed new interests, notably in diamonds and urinary calculi. The latter project was funded by the Medical Research Council and moved her interests towards medical science.

From the mid1950s Lonsdale was awarded many honors, both national and scientific. In 1956 she was created Dame Commander of the Order of the British Empire and the following year received the Davy Medal of the Royal Society, which she served as a Council member and Vice-President from 1960 to 1961. From 1959 to 1964 she was General Secretary of the British Association and served as President of the Physics Section during 1967. In 1968 she was its first female President. Eight British universities awarded her honorary degrees. Outside the UK she was vice-president of the International Union of Crystallography from 1960 to 1966, the first woman to hold this office. Lonsdaleite, a rare form of diamond found in meteorites, was named in her honor.

Lonsdale was also honored for her activities outside science, as a campaigner for improved conditions in prisons and an active member of the peace movement, both of which stemmed from her strong religious convictions. Both Kathleen and Thomas had become Quakers by convincement in 1935 and Kathleen regarded her roles as scientist, Quaker and mother as intimately connected. These activities were related through her experience of prison during World War Two when she was imprisoned in Holloway gaol for refusing to pay a fine of £2, imposed when she failed to register for employment and civil defence duties. There was no clause allowing for conscientious objectors and she was committed to gaol for a month. Her husband later claimed that this was the single most formative experience of her life and she later became involved in prison visiting, serving as a member of the Board of Visitors at Aylesbury Prison for Women and later as Deputy Chairman of the Board of Visitors, Bullwood Hall Borstal Institution for Girls. She was also an active member of the peace movement and joined the Atomic Scientists Association when it was first established, later serving as Vice-President. She attended several Pugwash meetings, was President of the British Section of the Women's International League for Peace and Freedom and a member of the East–West Committee of the Society of Friends. She travelled widely, sometimes managing to combine visits to prisons and discussions on global security with scientific meetings. Visits to the United States were sometimes problematic, with an Embassy official suggesting that this was because she had been to 'Russia, China and gaol'.

In 1961 Thomas retired and took on some of the burdens relating to her peace and prison work. They moved to Bexhill in Sussex and Kathleen added a substantial commute to her working day. She retired in 1968 and became an emeritus professor at University College, London. She continued to write on a range of topics, including urinary calculi and diamonds, to the end of her life.

Scientific Work

Lonsdale's first major work, carried out in conjunction with W.T. Astbury was on the relationship between X-ray diffraction patterns and the space groups from which they arose. This was the start of a longstanding interest in the production of tables to assist crystallographers in determining the structure of crystals. This was something she was able to continue with while working at home after returning from Leeds to London. In 1948 the newly established International Union of Crystallography selected her to chair its new Commission on Tables. Under her guidance new editions of the *International Tables* appeared in 1951 and 1959 and a further one was in preparation when she passed on the role in 1963.

Her second project, which came to fruition during her time at Leeds, was the structure of hexamethylbenzene, the first structure of an aromatic compound to be defined by X-ray diffraction. Her results showed that the benzene ring existed within the molecule as a planar hexagon, a conclusion that was confirmed by subsequent studies. This research was also methodologically important because of her successful application of Fourier methods to analyse X-ray diffraction patterns.

When she returned to experimental work at the RI her interests shifted to optical and magnetic anisotropy as an aid to structural analysis. During the late 1930s she developed an interest in the thermal movement of atoms in crystals. This involved the development of new experimental methods in order to take measurements at low temperatures. Diamond presented particular problems to Lonsdale and her fellow researchers and she continued her work after her move to University College, where she also worked on diamonds with H.J. Milledge (formerly Greville-Walsh) who first came to work with her in 1949 and became an important collaborator for the remainder of her career. Lonsdale continued to work on lattice dynamics for the rest of her life and left an unfinished manuscript on the thermal expansion of crystals when she died.

In her later years Lonsdale turned her attention to problems of medical and biological interest. Her first foray into this field started in 1954 when she was alerted to the possibility of links between the pharmacological activity and the geometrical structure of the *n*-methonium compounds. This was following during the early 1960s by work on urinary calculi. By studying the structure and composition of these stones it was hoped to learn something about their formation and use this knowledge to identify a mechanism for inhibiting their formation and growth.

Lonsdale's researches were united by her innovative and methodical approach to experimentation and her ability to apply mathematics to her data. She worked closely with her laboratory technicians at both the RI and University College, and included them as coauthors on many of her publications. Her students benefitted from her critical attention to their work and researchers came from around the globe to work with her.

Kathleen Lonsdale was a scientist of distinction who was also active as a campaigner for prison reform and in the peace movement. Her scientific work in the field of crystallography brought many accolades and she was one of the first women to be elected to the Royal Society as well as the first to serve as President of the British Association for the Advancement of Science. Her work for prison reform and as a peace activist was under-pinned by her strong religious convictions. Unusually for women scientists of her generation, she was able to combine her scientific work with motherhood. This was only possible through the strong support of her husband, Thomas. Without his willingness to accept the unusual domestic arrangements that a working wife entailed it would have been very difficult indeed for Kathleen to have continued her scientific work while her children were small, and her long and successful career would have been cut short.

Literature

Baldwin, M. (2009) Where are your intelligent mothers to come from?: marriage and family in the scientific career of Dame Kathleen Lonsdale (1903–71). *Notes and Records of the Royal Society* **63**, 81–94.

Childs, P. (2003) Woman of Substance at http://www.rsc.org/chemistryworld/ Issues/2003/January/substance.asp (accessed 28 July 2010).

Hodgkin, D.M.C. (1975) Kathleen Lonsdale. *Biographical Memoirs of Fellows of the Royal Society*, **21**, 447–484.

Hudson, G. (2004) Lonsdale, Dame Kathleen (1903–1971). *Oxford Dictionary of National Biography*, Oxford University Press, Sept 2004; online edn, Oct 2009 http://www.oxforddnb.com/view/article/31376 (accessed 2 Aug 2010).

Julian, M.M. (1995) Kathleen and Thomas Lonsdale: forty-three years of spiritual and scientific life together. in *Creative Couples in the Sciences* (eds H.M. Pycior, N.G Slack and P.G. Abir-Am), Rutgers University Press, New Brunswick, pp. 170–181.

Lonsdale, K. (1970) Women in science: reminiscences and reflections. *Impact of Science on Society*, **20**, pp. 54–55.

Lonsdale, K. (1964) *I believe...* Cambridge University Press, Cambridge

Rayner-Canham, M and Rayner-Canham, G. (2008) *Chemistry Was Their Life: Pioneer British Women Chemists, 1880–1949*, Imperial College Press, London.

Marthe Louise Vogt (1903–2003)

Annette B. Vogt

Marthe L. Vogt was a German-British physician, chemist and pharmacologist. She contributed fundamental research in neuropharmacology and was one of the leading neuroscientists of the 20th century. She made famous discoveries on the anatomy and the distribution of neurotransmitter and adrenal hormones. In 1952 she became a Fellow of the Royal Society.

Marthe L. Vogt was born in Berlin on September 8, 1903. She grew up in the family of the famous brain researchers, Cécile (1875–1962) and Oskar (1870–1959) Vogt. Her younger sister Marguerite Vogt (1913–2007) became a geneticist and cancer researcher.

After a high school education, she studied medicine and chemistry from 1922 until 1927 at the Berlin University, the Friedrich-Wilhelms-Universität, and she received – very unusual in the 20th century – two doctoral degrees. First, in 1928 she received her doctoral degree in medicine with a thesis on brain research ("Über

Marthe Louise Vogt

European Women in Chemistry. Edited by Jan Apotheker and Livia Simon Sarkadi
Copyright © 2011 WILEY-VCH Verlag GmbH & Co. KGaA, Weinheim
ISBN 978-3-527-32956-4

omnilaminäre Strukturdifferenzen und lineare Grenzen der architektonischen Felder der hinteren Zentralwindung des Menschen") which was published in the journal her parents edited. Only one and a half years later, in 1929, she received her second doctoral degree with a thesis on biochemistry ("Untersuchungen über Bildung und Verhalten einiger biologisch wichtiger Substanzen aus der Dreikohlenstoffreihe") which was published in the journal *Biochemische Zeitschrift*, edited by her doctor father, the famous biochemist Carl Neuberg (1877–1956). She was employed as a doctoral student in his Kaiser Wilhelm Institute for Biochemistry between 1927 and 1929. Thanks to her family background (her mother was French) she learned French, and because of the scientific relations of her parents with Russian-Soviet colleagues she also learned Russian.

From December 1930 until April 1935 Marthe L. Vogt worked in the Kaiser Wilhelm Institute (KWI) for Brain Research in Berlin-Buch, directed by her parents Cécile and Oskar Vogt. It was the only KWI which was lead by a couple in science. Furthermore, the KWI for Brain Research was the only one where many women scientists, including married women scientists, were employed and got career chances. In June 1931, Marthe L. Vogt became the head of the small chemical department where she investigated the reactions of certain chemical substances in the brain. This research was carried out partly in competition with the guest department in the KWI for Psychiatry in Munich, lead by the US researcher Irvine H.

Marthe Louise Vogt (Archive of Berlin-Brandenburg Academy of Science, Berlin: Abt. Sammlungen, Foto-Sammlungen, Marthe Vogt.

Page (1901–1991). Both departments contributed to the field which was later called neurochemistry.

Because of the Nazi regime in Germany Marthe L. Vogt decided not to live under such circumstances and became an emigrée. In 1935 she travelled to Great Britain thanks to a fellowship from the Rockefeller Foundation. In 1939 she asked for British citizenship which she finally received in 1947. In Great Britain, first, she worked in London from 1935 until 1936 in the laboratory F4 of the National Institute for Medical Research under Sir Henry H. Dale (1875–1968) in Hampstead/ London. Then she studied again, and in 1937 she received her third academic degree, the PhD at Cambridge University. Thus began one of the rare "success stories" of a German emigrée. From 1937 until 1940 she was Fellow at Girton College Cambridge, one of the old and famous Women's Colleges, thanks to the Alfred Yarrow Research Fellowship. From 1941 until 1946 she was employed at the British Pharmaceutical Society where she did research on pharmacological problems. From 1947 until 1960 she taught and did research at the Pharmacological Laboratory of the University of Edinburgh. In 1960, she moved back to Cambridge and became head of the Pharmacological Unit of the Agricultural Research Council Institute of Animal Physiology. After her retirement in 1968 she continued her research until the late 1980s, when she moved to La Jolla, California, to stay together with her sister Marguerite Vogt who was working at Caltech as a cancer researcher. Here, Marthe L. Vogt died the day after her 100th birthday, on September 9, 2003.

The scientific work of Marthe L. Vogt was related first to the study of chemical substances and their effects on the brain. Together with Sir Henry H. Dale and Wilhelm Feldberg she investigated the role of chemical neurotransmitters, in 1936 they published their article. Her most famous publication was published in 1954 "The concentration of sympathin in different parts of the Central Nervous System under normal conditions and after the administration of drugs", and she became one of the leading neuroscientists. She made important contributions to the understanding of the role of neurotransmitters in the brain. Marthe Louise Vogt was one of the most important and famous women scientists in the 20th century. She was one of the pioneers of neuropharmacology and neuroendocrinology and one of the leading neuroscientists of the 20th century. Among her disciples Susan Greenfield is one of the most prominent.

Marthe Louise Vogt was well acknowledged in the scientific community. In 1952 she was elected a Fellow of the Royal Society. Later. she held several honorary doctorates. She became a lifelong Fellow of Girton College, Cambridge in 1960, in 1977 she was elected a Foreign Honorary Member of the American Academy of Arts and Science. She was awarded several Prizes and Medals: in 1974 she got the Schmiedeberg Plakette of the German Society for Pharmacology and Toxicology; in 1976 she was awarded the Thudichum Medal of the Neurochemical Group of the British Biochemical Society; in 1981 she earned the Royal Society Gold Medal ("Queen's Gold Medal"), and in 1983 the Wellcome Gold Medal of the British Pharmacological Society. She was a Member of the British Pharmacological Society, the Hungarian Academy of Sciences, the British Association of Psychopharmacology, an Honorary Fellow of the Royal Society of Medicine, an Honorary Member of the Physiological Society, to mention only some of them.

Literature

Important publications:

Biographisches Handbuch der deutschsprachigen Emigration nach 1933 (International Biographical Dictionary of Central European Emigrées 1933–1945) (1983) Vol. II, 2, p. 1195 (eds. Röder, W. and Strauss, H.A.) Saur Verlag, München.

Greenfield, S. (1993) Marthe Louis Vogt F.R.S. (1903–) in *Women Physiologists* (eds L. Bindman, A. Brading, T. Tansey) Portland Press, London, Chapel Hill, pp. 49–59.

Mason, J. (1995) The women fellows' jubilee, *Notes and Records R. Soc. (London)*, **49** (1), 125–140.

Mason, J. (1992) The admission of the first women to the Royal Society of London, *Notes and Records R. Soc. (London)*, **46** (2), 279–300.

Medawar, J. and Pyke D. (2001) *Hitler's Gift. The True Story of the Scientists expelled by the Nazi Regime*, Foreword by Dr. Max Perutz, Arcade Publishing, New York.

Ogilvie, M. and Harvey, J. (Eds) (2000) *The Biographical Dictionary of Women in Science.*

Pioneering Lives from Ancient Times to the Mid-20th Century, Vol. 2, Routledge, New York and London, pp. 1330–1331.

Vogt, A. (2008) *Wissenschaftlerinnen in Kaiser-Wilhelm-Instituten. A-Z.* 2. erw. Aufl., (= Veröffentlichungen aus dem Archiv zur Geschichte der Max-Planck-Gesellschaft, Bd. 12), Berlin, pp. 200–204.

Vogt, A. (2007) *Vom Hintereingang zum Hauptportal? Lise Meitner und ihre Kolleginnen an der Berliner Universität und in der Kaiser-Wilhelm-Gesellschaft*, Franz Steiner Verlag, Pallas&Athene, Stuttgart, Vol. 17.

Vogt, M. (1954) The concentration of sympathin in different parts of the central nervous system under normal conditions and after the administration of drugs, *J. Physiol.*, **123**, 451–481.

Vogt, M. (1947) Cortical lipids of the normal and denervated suprarenal gland under conditions of stress, *J. Physiol.*, **106**, 394.

Vogt, M., Dale, H.H. and Feldberg, W. (1936) Release of acetylcholine at voluntary nerve ending, *J. Physiol.*, **86**, 353–379.

Carolina Henriette MacGillavry (1904–1993)

Mineke Bosch

■ Caroline MacGillavry was a chemical crystallographer and one of the pioneers of X-ray diffraction of crystals in the Netherlands. Highly gifted in mathematics, she developed independently and at the same time as two American scientists, the Direct Method in calculating X-ray crystallographic data. Her dissertation was on conjoined crystals, from 1950 she was involved in the study of Vitamin A. She became one of the first women professors in the Netherlands and the first woman to be appointed as a member of the Royal Netherlands Academy of Arts and Sciences (Koniklijke Nederlandse Akademie van Wetenschappen (KNAW)). After she met the graphic artist M.C. Escher, she became interested in his intriguing work with 'impossible images', and, in 1965, published the book *Symmetry Aspects of M.C. Escher's Periodic Drawings.*

Carolina Henriette MacGillavry, Lien to family and friends, Mac to her coworkers and colleagues, was born as the second of six children in an intellectual family environment. Her father was a brain surgeon, her mother a teacher. She attended the Barlaeus Gymnasium in Amsterdam, and started studying chemistry at the University of Amsterdam in 1921. She was a talented mathematician and was highly intrigued by the discovery of quantum mechanics in 1925. As a student she once gave

C. H. MacGilavry, photo from KNAW.

European Women in Chemistry. Edited by Jan Apotheker and Livia Simon Sarkadi
Copyright © 2011 WILEY-VCH Verlag GmbH & Co. KGaA, Weinheim
ISBN 978-3-527-32956-4

a presentation on the quantum mechanical computations of the H-molecule. En-
thused by the senior lecturer J.M. Bijvoet, who was a pioneer of the X-ray diffraction of crystals, she specialized in solid-state chemistry, and especially chemical
crystallography. She earned her master's degree in 1932 cum laude, and was assistant to Professor A. Smits at the Laboratory for General and Inorganic Chemistry
at the University of Amsterdam. Working with X-ray diffraction, she became fascinated by the beauty of crystals that resulted from exactly the right interplay between
order, regularity, boundaries, color, variation, natural irregularities, and deviations
in the crystalline structure. In January, 1937 she received her PhD, again cum
laude, with a dissertation on the structure of several conjoined crystals. Immediately after earning her doctorate, MacGillavry worked for six months as assistant at
the Physical Chemistry Laboratory in Leiden. By September of the same year she
took up a position as assistant to Bijvoet in the Laboratory for Crystallography at the
University of Amsterdam, where she was to work the rest of her career. With Bijvoet she published the 1938 reference work *Röntgenanalyse van Kristallen* (X-ray
Analysis of Crystals) and several important contributions to the Dutch journal of
physics (*Nederlands Tijdschrift voor Natuurkunde*), to *Nature* and to the Dutch chemical journal *Chemisch Weekblad*. In 1941 she was appointed as curator of the Laboratory, which was followed in 1946 by her appointment as senior lecturer in chemical crystallography at the University of Amsterdam.

After the end of the Second World War, MacGillavry ventured into new research
areas. She also broadened her scope internationally. In 1947 she attended the first
European conference on crystallography, organized by W.L. Bragg at the Royal Institution in London. In 1948/49 she went with a UNESCO–American Chemical Society fellowship for a period of several months to the United States. While there she
represented the Netherlands at the first congress of the International Union of
Crystallography. At this conference she was elected as a member of the committee
on the three-volume reference work *International Tables for X-ray Crystallography* of
the International Union of Crystallography. She was one of the editors of the third
volume, *Physical and Chemical Tables*, that was published for the first time in 1962.
Co-editors were the British crystallographer and friend Kathleen Lonsdale and the
Dutch Gerard Rieck.

In the USA MacGillavry used an analog computer that was especially developed
to perform crystallographic research, eliminating the need for a lot of manual calculations. In the same year, at a meeting of the Crystallographic Society of America (a forerunner of the American Association of Crystallography) she presented her
work on the 'Direct Method' of computing in crystallography. This was fully recognized as an independent and further elaboration of the D. Harker and J.S. Kasper
equation. Her contributions to the method were published by Pepinski in his *Computing Methods and the Phase Problem in X-ray Crystal Analysis* in 1952. Between
1954 and 1960 MacGillavry was an active member of the executive committee of the
International Union of Crystallography.

Her extensive knowledge, the subjects she chose, and her control of the experimental techniques gained MacGillavry a leading role in the national and international community of X-ray analysts. In 1948 she was one of the initiators of

FOMRE, the Dutch foundation for fundamental research of matter with electronic and X-ray diffraction, which she presided over until 1972. She contacted and worked with the researchers in the National Laboratory of Philips, who were interested in her efforts to get a large computer. All these activities resulted, in 1950, in her appointment as extraordinary professor in chemical crystallography at the University of Amsterdam. One can wonder whether she would have been appointed full professor if she had been a man, but at least full recognition of her excellence came in 1958. In the meantime she was the first woman to be elected a member of the Royal Netherlands Academy of Arts and Sciences (KNAW) in 1950. In 1961 she became director of the Laboratory for Crystallography in Amsterdam, which by then had become an independent institution. In 1966 she formed a life partnership with a medical specialist, J.H. Nieuwenhuijzen.

Crystallographers that shared the pioneering phase had a strong sense of urgency of what they are doing, which fostered (international) cooperation rather than competitiveness. In her inaugural lecture in 1950 she expressed some of this sentiment by characterizing crystallography and its practitioners as an attractive research area also for women. Scientists needed to possess character traits such as an artistic nature, mental elasticity, spatial imagination and intuition. Moreover, crystallography was practised in small groups rather than in large chaotic natural sciences and chemical laboratories that were practically male households. Her appearance surprised the larger public that still equated women scientists with the inelegant drudge. It earned her favorable press attention and observations of her as youthful, easy-mannered 'dressed in a comfortable blouse and a gay checked skirt'. In the laboratory MacGillavry managed to create a congenial atmosphere in which many work relations turned into friendships. From 1957, 21 researchers did their PhD under her supervision. There was also an elaborate laboratory culture in which successes were celebrated and accompanied by music.

After the 1950s, her structural research became dominated by the study of vitamin A and related compounds. This research was important for the physiological process of sight. It was during these specialized activities that she started to combine her love of crystallography with that of art and nature. An important aspect of this was her meeting with the graphic artist M.C. Escher in 1959. She immediately saw that the recurring principles of the forms in his work were akin to regularities in crystalline material. She asked him to exhibit his work at the occasion of the congress of the International Union of Crystallographers in Cambridge in 1960. At the request of the Union she wrote a monograph on his work, which was published in 1965 as *Symmetry Aspects of M.C. Escher's Periodic Drawings*. However, it was primarily after her retirement as a professor that she started to get involved in the historical, mathematical, and artistic aspects of crystallography that reappeared in other settings such as art. Based on publications and lectures about symmetry in things like mirrors, animals, crops, minerals, and the Dutch polder landscape, she was able to explain works of art, as well as nature and science, to a broad public. The audience at the lecture at the Amsterdam conference of the International Union of Crystallography in Amsterdam in 1975, with the title 'Order and Beauty', for instance, learned how crystals get their appearance, how order and irregularities create beauty in all the areas referred to above, and how great parallels and analogies can be observed between these very different domains.

"Pure science and its social impact, those are two different domains that can be separated widely", said MacGillavry in her farewell lecture as Professor of Chemical Crystallography at the University of Amsterdam. Popularization of science might bridge that gap, but few scientists succeeded in doing that. There is no doubt, however, that MacGillavry herself managed to do that from the 1960s, based on her vast knowledge and multifaceted interest. Within her specialized field, she focused on a very diverse range of research subjects, and she managed to be innovative in several directions. She also possessed extensive knowledge of plants and animals, as well as of literature, art and music, and found time to play violin in the laboratory quartet. Her public lectures drew large audiences, attracting both scientists and nonscientists. Through her ability to express, in concrete terms, what the beauty of crystals, for example, was based on, she succeeded in bringing an understanding of pure scientific knowledge a lot closer to home for many people by the time of her death in 1993.

After her death Caroline MacGillavry was buried next to her partner who had died seven years before her. During her life she earned several decorations and honorary memberships. She left a substantial part of her inheritance to the KNAW in order to install a MacGillavry Foundation which aims at supporting young researchers who with their natural science research contribute to the solution of problems of the developing countries.

Literature

Bosch, M. (2006) Fascinated by Crystals' Sublime Beauty. Carolina Henriette MacGillavry, First Woman of the Royal Netherlands Academy of Arts and Sciences (KNAW), in R. Oldenziel and M. Bosch (eds) *Curious Careers. An Unexpected History of Women in Science and Technology*, Stichting Historie en Techniek, Eindhoven.

Bruinvels-Bakker, M. and de Knecht-van Eekelen, A. (1997) Carolina H. MacGillavry: eerste vrouw in de Koninklijke Nederlandse Akademie van Wetenschappen; over de schoonheid van kristallen, vrouwelijke intuïtie en lenigheid van geest. *Gewina*, **20**, 309–331.

Bruinvels-Bakker, M.Th. Mac Gillavry, Carolina Henriette (1904–1993), in *Biografisch Woordenboek van Nederland*. URL: http://www.inghist.nl/Onderzoek/Projecten/BWN/lemmata/bwn5/mac_gillavry (accessed 13-03-2008).

Looijenga-Vos, A. (1994) Carolina Henriëtte Mac Gillavry, in *Levensberichten en herdenkingen 1993. Koninklijke Nederlandse Akademie van Wetenschappen*, Amsterdam. pp. 54–59.

Lucia de Brouckère (1904–1982)

Brigitte van Tiggelen

Lucia de Brouckère was appointed as an assistant to Professor J. Timmermans (1899–1986) in 1927. Immediately after the usual congratulations from the rector of the Université Libre de Bruxelles (ULB), she was advised not to expect anything further. The future proved the rector's reservations wrong, since she became the first woman appointed Professor of Chemistry in Belgium, and also the first female professor in a Belgian science faculty. She was a remarkable teacher, educating hundreds of students in general chemistry and mentoring many specialized chemists. She was also a very deeply committed socialist and free thinker who engaged in numerous social activities and associations.

Lucia Florence Charlotte was born on July 13 in 1904 in Brussels. Her father, Louis de Brouckère was a famous politician and member of the socialist party; he even went to prison for having published an antimilitarist pamphlet in 1898. Despite his views, he enlisted in the Belgian army in 1914, aged 44. After the defeat, when Lucia was 10 years old, the family fled to England. Louis' successful political career led him to many posts in Belgium and also to represent his country in several international posts, such as the League of Nations. He was also an early advocate for women's education, and Lucia's mother, Gertrude Guïnsburg, had herself studied, which was most unusual at that time. It is thus no surprise that Lucia was encouraged to enter university after having attended school in Belgium and in Great

Lucia de Brouckère, photo provided by author.

European Women in Chemistry. Edited by Jan Apotheker and Livia Simon Sarkadi
Copyright © 2011 WILEY-VCH Verlag GmbH & Co. KGaA, Weinheim
ISBN 978-3-527-32956-4

Britain during the First World War. This was no easy task for Latin and Greek were required subjects to enter university and those were not taught in girl schools at that time.

Lucia was deeply impressed by her father's social commitment and revered his memory her whole life long. Her own commitment surely stems from this early exposure to his ideals. But she did not want to follow in her father's footsteps completely. She never aimed at a political career and, though very much attracted to history, she opted for sciences. She thought she lacked any literary talent but had the taste for logical reasoning needed to succeed in the sciences. She knew she would enter any career with ease with the support of her father after a graduation in "Philosophie et Lettres", but this was precisely what she did not want to do. She wanted to owe her success solely to her own work. In her choice of chemistry, came another ingredient, the admiration for a woman, Daisy Verhogen, who was "chef de travaux" in organic chemistry at the ULB (this is a higher rank than assistant in a scientific career without ever leading to any academic title such as professor). Significantly, little is known about Daisy Verhogen. History still needs to investigate such secondary figures, who paved the way for the later women in science.

In 1927 she defended her thesis "L'adsorption des electrolytes par les surfaces cristallines", for which she received a prize from the Académie Royale de Belgique. Her whole research focused on the adsorption of metallic ions by crystalline surfaces, using four different experimental techniques to compare the results. Precision and strictness are the signature of her experimental work. Among other things she demonstrated that the triiodide ion is fixed perpendicular to the surface of barium sulfate, which brought her into controversy with the Dutch originated American chemist I.M. Kolthoff (1894–1993). The divergence was solved when the two chemists exchanged their samples and found out that they were different due to different methods of preparation. She became "Agrégée de l'enseignement supérieur" in 1933 (this is a title required to teach at university level), and acted as a relief teacher for a while. In 1937, she received a lectureship at ULB for general chemistry, which made her the first woman to teach in a Belgian science faculty. Her research focus shifted with the war. Moving with her father to England in 1914, she studied the corrosion of lead by tap water and the corrosion of aluminum–magnesium alloys by atmospheric air, but also more confidential matters such as the corrosion due to desert sand of the copper contacts of a tank horn! She also led the Industrial chemistry section at the Ministry for economic affairs of the Belgian government exiled in London.

As soon as she returned to Belgium in 1944, she reorganized the chemistry curriculum and found financial support to invite foreign professors. In 1946, she was appointed Director of the General Chemistry Laboratory and in 1951, Director of the Mineral and Analytical Chemistry Laboratory, identified by her team as "La patronne". She became interested in colloids and macromolecules, of which she investigated the physico-chemical properties when in solution. During several years, she collaborated with Ilya Prigogine on joint research on thermodynamics in the liquid phase, providing the experimental results while he was responsible for the theoretical developments.

The Prix Wetrems was granted in 1953 by the Académie Royale de Belgique to honor her scientific work. Her research slowed in the 1960s, as she taught more and more courses. She was appointed full professor in 1945 (retrospectively for 1942 – during the second war the ULB was indeed closed by German occupation) and was given the responsibility of teaching general chemistry. She designed a whole new course that was based more on the fundamental principles and reasoning than memory, as was usually the case for beginners courses, including, nevertheless, numerous original experimental demonstrations. She successively accepted other teaching duties, ranging from analytical chemistry, to colloids and macromolecules to physical or inorganic chemistry. She was also involved from 1939 in the training of future science teachers. The attention to basic concepts and the experience gained through teaching them allowed her to write a historical book : *Evolution de la Pensée Scientifique : Evolution des Notions d'Atome et d'Elément* (Brussels, 1982) – (Evolution of Scientific Thought: the Evolution of the Ideas of Atoms and Elements).

Within the Science Faculty she served as President in 1962–1963, and, in 1965, joined the board of the International Solvay Institutes of Physics and Chemistry which are attached to the ULB. When the University underwent a vibrant student call for reform in 1968, she acted as President in the Assembly working on ULB's new statutes, which showed the extent of the moral authority gained among both her colleagues and the students. She also sat on the board of many scientific affairs related commissions, inside and outside the University. Always curious about new trends in teaching and communicating sciences, she participated in the foundation of the association *Jeunesses Scientifiques de Belgique* in 1957, which promoted scientific activities for young people and is still very much active to day. Her dedication to the younger generation also showed in the Fondation Lucia de Brouckère, created on the occasion of her retirement to allow young chemistry students from ULB to stay abroad.

Her dress was severe: she would invariably wear a dark or grey suit, when she was not running around wearing a lab. coat. But behind this austere image, everyone involved with her would soon discover the enthusiastic, meticulous and hard working personality. Full of energy, generous, socially aware, always attending meetings well prepared, she served her ideal on every possible occasion. She was always passionate about communicating her knowledge and science. Her personality, as well as her pedagogical talents and her research skills impressed thousands of students over the years until she retired in 1974. Her reputation however was even greater in the socio-political realm. In 1934, she was chosen as the first chairwoman of the Women's World Committee Against War and Fascism and, true to her conviction, she campaigned for Republican Spain in 1936.

As a free-thinker, she was initiated as a freemason in the Lodge *La Sincérité et la Paix Réunies* of which she became Worshipful Master between 1964 and 1966. Engaged in many free-thinkers' circles she realized that their efficiency was hindered by their dispersion and, in 1969, took the initiative of a *Centre d'action laïque* to coordinate all efforts of these specialized and scattered organizations, especially when speaking to the public authorities.

Being the first female academic teacher, she was often interviewed about the equal opportunity question. She never joined in any feminist group. She was convinced that the liberation of women could not be achieved without the support of men, or outside existing institutions. Therefore one had to keep acting inside mixed groups at every level of society, and also accept challenge or reservation, as was the case when she became full professor in 1945: she had to promise she would resign in two weeks if ever there were some authority problem with an audience of several hundred students! She considered things were improving, conceding, however, that at equal proficiency, men retained an advantage over women. The struggle was not over yet, and, in that respect, the private sector was even slower to accept women in higher positions than the academic sector.

Her main concern though was to educate the younger generation, boys and girls. In her mind, university should not produce highly skilled specialists but individuals trained to apply the scientific method and able to adapt themselves their whole life long: "What is needed are men and women who have the courage to remain students until their last breath, men and women who take every opportunity to improve or correct their knowledge" (address to students in 1960). And while stressing the danger, the uncertainty or the distress suffered by a growing number of people in society, she would also speak confidently of the social role and responsibility the next generation of scientists would have to take on. This message is still timely and there is no doubt that, now as then, an apprenticeship in chemistry serves Lucia de Brouckère's perspective particularly well.

Literature

Nasielski, J. (2007) de Brouckère, Lucia, in *Nouvelle Biographie Nationale*, Academie Royale de Belgique, t. IX, pp. 111– 114.

van de Vijver, G. and Lemaire, J. (1993) *Science et Libre Examen : un Homage à Lucia de Brouckère*, Espace de Libérté/CAL, Brussels.

van Tiggelen, B. (2004) Lucia, dite Lucie, de Brouckère (1904–1982), in *Chimie et Chimistes de Belgique*, Labor, Brussels, pp. 88–89.

Berta Karlik (1904–1990)

Maria Rentetzi

■ Berta Karlik is known for the discovery, in 1943, – in collaboration with Traude Cless-Bernert – of the natural occurrence of isotopes of astatine by observation of their radioactive alpha particle decays. Two years later she resumed her duties as Director of the Institute for Radium Research in Vienna. In addition, in 1956, she was promoted to the position of full professor, the first woman in Austria in such a position. Throughout her career, Karlik was honored by several awards. In 1973 the Austrian Academy of Sciences elected her as a member, the second female member of the Academy after Lise Meitner. She was also a founding member of the Austrian Physical Society and was among those who strongly initiated the Austrian membership of CERN. Politically aware, Karlik heartily supported those of her Jewish colleagues who were persecuted during the Nazi period, while, after the war, she became active in the Austrian Association of University Women.

Karlik was born in 1904 in an upper class Viennese family. Her father, Carl Karlik, was director of the national mortgage institution for Lower Austria and Burgenland. She lived in a small castle in Mauer, a Viennese suburb. Adopting the status of her class, she received her primary school education at home, learned to play piano and to speak several languages while also taking classes in painting. From 1919 to 1923 she attended the Reform-Realgymnasium in the thirteen district of Vienna and during the academic year 1923/24 she was registered as a regular student at the Philosophical Faculty of the University of Vienna. In 1927 Karlik defended her thesis to Stefan Meyer, Director of the Institute for Radium Research in Vienna and to Hans Thirring, Director of the Institute for Theoretical Physics. In the meanwhile she had become an essential member of Hans Pettersson's research group at the Radium Institute, focusing especially on the scintillation counter. The same year Karlik completed also the examination for the teaching profession and accepted a position at a *Realgymnasium* in Vienna.

A fellowship from the International Federation of University Women allowed Karlik to spend some time away from Vienna's Radium Institute. During the academic year 1930/31, she moved to William Bragg's laboratory in London. Her research interests were centered on crystallography and the use of X-rays in the

European Women in Chemistry. Edited by Jan Apotheker and Livia Simon Sarkadi
Copyright © 2011 WILEY-VCH Verlag GmbH & Co. KGaA, Weinheim
ISBN 978-3-527-32956-4

Berta Karlik working on the scintillation method at the Radium
Institute. (Source: Agnes Rodhe personal archive.)

study of the structure of crystals. It was her knowledge of radiophysics that Karlik
brought to Bragg's laboratory, forming a group with the crystallographers Ellie
Knaggs and Helen Gilchrist. The same year she also visited also Marie Curie's
laboratory in Paris. Back in Austria Karlik teamed up with physicist Elizabeth Rona
on the study of the ranges of alpha particles emitted from actinium and polonium.

Around the same time Karlik joined a group on seawater research, initiated by
the Swedish physicist Hans Pettersson. On the border zone of oceanography and
radioactivity, Karlik and Friedrich Hernegger, a research student at Vienna's Radium Institute, brought up concerns on biological issues in relation to the uranium
content of seawater. During the Second World War Karlik reached the climax of her
research. Together with Traude Cless-Bernert, a research student at Vienna's Radium Institute, she proved the existence in nature of the element with atomic number 85, that is, astatine. For this Karlik received the Haitinger-Preis for Chemistry
from the Austrian Academy of Sciences in 1947.

Karlik's university career started in 1937 when she received the *Venia Legendi*, the
formal requirement for the right to teach at the university and became a *Dozentin*.
Three years later she was named assistant and in 1942 *Diätendozentin*. Right after
the end of the Second World War Karlik resumed the Directorship of the Institute
for Radium Research and initiated its restructuring and renovation. She arranged
the installation of a Cockroft–Walton accelerator, progressing physics research
from the electromagnets and accumulators of the 1920s to the large accelerators of
the 1960s. In 1950 she became *ausserordentliche Professorin* at the University of Vienna and the first woman *ordentliche Professorin* in 1956. She retired in 1974 having
contributed most to the advancement of nuclear physics research in Austria. She
continued to work until her death on February 4, 1990 in Vienna at the age of 86
years.

Karlik's dissertation topic was on the dependence of scintillations occurring by charged particles striking zinc sulfide and the nature of the scintillation process, a cutting edge topic at the time in research on radioactivity. The scintillation technique used for the detection of nuclear particles lay at the center of a major scientific controversy between two research groups, that of Ernest Rutherford at the Cavendish Laboratory in Cambridge and that of Hans Pettersson at the Institute for Radium Research in Vienna. Karlik played an essential role throughout this heated debate that took place during the 1920s.

In its generic form the scintillation counter was a very simple instrument. It consisted of a screen, a thin glass plate spread with an equally thin layer of zinc sulfide. When it was struck by charged particles, the screen produced light flashes. The scintillations were observed through a microscope, which was specifically designed to increase the brightness of the flashes. By manipulating the microscope and its light-gathering power, the experimenter could work with weak radioactive sources and still observe a fair number of particles. The observations carried out in a dark room were tiring and tiresome and the counting fragile, heavily dependent on the experience of the observer.

In her earlier work, Karlik described a photometric method for determining the relation between the range of alpha particles and the brightness of the scintillations for differently prepared zinc sulfide screens. In her model of the counter, she introduced photographic plates. In order to reduce the light entering the eye through the microscope she placed photographic plates between the objective and the eyepiece. In collaboration with another female physicist, Elizabeth Kara-Michailova, Karlik also measured the luminescence produced by alpha particles emitted from polonium by means of the photoelectric current of a rubidium cell. The introduction of the cell was an innovation towards a mechanized, and thus considerably more objective, way of recording scintillations. In their works that followed, besides discussing the experimental details of the relation between the brightness of the scintillation and the energy given up from the alpha particles of the source, the two women suggested a theoretical hypothesis to explain the mechanism of the scintillation process. They were concerned with more than manipulating the instrument, preparing and gauging the scintillation screens, and experimenting with several different elements. They went one step further, suggesting that the zinc sulfide possesses distinct points that are already in an active condition before they are struck by the particles.

However, there was a considerable difference between the results obtained by Karlik and similar experiments done in Cambridge concerning the question of how the amount of light entering the eye from an individual scintillation affects the number of total scintillations observed in the disintegration process of light elements. To resolve the discrepancies James Chadwick, Rutherford's collaborator, visited Vienna in 1927. There he was able to show empirically that the Viennese research group was wrong in the number of scintillations they were claiming to count by repeating the controversial experiment and asking the women of the group to do the counting of the scintillations. As Chadwick described his own visit "I arranged that the girls should count and that I should determine the order of counts. I made no change whatever in the apparatus, but I ran them [the counters] up and down the scale like a cat on a piano...' Karlik was not one of the counters but stood in the audience. "The younger ones..." affirmed Chadwick "stand around stiff-legged and with bristling hair". Yet all of these women used as counters by Chadwick had come a long way in designing their own instruments and experiments, playing an instrumental role in Pettersson's research group.

With this kind of appreciation of women's work in science Karlik, having played a central role in improving the counter, abandoned the technique as the center of her research focus. Instead she accepted the Crosby–Hall stipendium from the International Federation of University Women in 1930 and moved to Britain for a year. She was also able to visit Marie Curie's laboratory in Paris along with the Pasteur Institute and Louis de Broglie's laboratory. According to Otto Hahn, "in our days it is a great advantage to Karlik that she worked next to prominent scientists both in England and France and later in Sweden. Because of that she was able to broaden widely her horizon more than is possible in a normal life in the sciences". As Karlik's later career indicates Hahn was proved right.

Karlik belonged to a generation of physicists who were immersed in the everyday world, socially active, politically engaged and culturally informed. During the interwar period she was a member of a circle of some young Austrians with interests in music and democratic politics. As she later confessed in a radio interview "all I can say is that I have very versatile intellectual interests, very versatile. Therefore, I am not oriented only to physics and the sciences. I am interested in questions about art and history...I am also interested in music". Physics, however, played a fundamental role in her life. During the Nazi period Karlik's status at the Radium Institute was downgraded. Struggling to retain her own research position, with a strong feeling of ambivalence, "torn to pieces", Karlik, who had an opportunity to leave the country, decided to stay in Vienna. As she admitted to Hellen Gleditsch, "I think perhaps some of my English friends wonder why I am not leaving Germany in protest. I have come to the conclusion that protest on the part of a German individual is quite useless at the moment and that more is done by staying and trying to improve matters from within the country". In the long run Karlik was indeed able to improve the country, not only by improving nuclear physics research in Austria but also by standing by her Jewish friends and colleagues during the political instability of the war period.

Literature

Bischof, B. (2004) *"Junge Wienerinnen Zertrümmern Atome..." Physikerinnen am Wiener Institut für Radiumforschung*, Talhheimer Verlag.

Lintner, K. (1990) *Berta Karlik, Nachruf*, Österreichischen Academie der Wissenschaften, Wien.

Rentetzi, M. (2008) *Trafficking Materials and Gendered Experimental Practices: Radium Research in Early Twentieth Century Vienna*, Columbia University Press, New York.

Elsie May Widdowson (1906–2000)

Sally Horrocks

■ Elsie Widdowson will be best remembered for her scientific partnership with Robert McCance, widely recognised for its contributions to the detailed knowledge of the nutritional composition of foods. This underpinned the publication in 1940 of *The Chemical Composition of Foods* which was regularly updated thereafter and is now known as *McCance and Widdowson's The Chemical Composition of Foods*. The volume rapidly became an essential resource for subsequent generations of nutrition scientists and others interested in the subject around the globe. She also made significant contributions to practical investigations of the nutrition and health of communities under stress. In addition her research encompassed neonatology and the feeding of infants. Formal recognition of her scientific achievements, including election as FRS in 1976, CBE in 1979 and Companion of Honour in 1993, did not come until after formal retirement in 1973.

Elsie May Widdowson was born on 21 October 1906 in Wallington Surrey, the elder of two daughters of Thomas Henry Widdowson, a grocer's assistant and his wife Rose Elphick. Both she and her younger sister, Ethel Eva, gained scholarships to the Sydenham county secondary school. Here they found encouragement from staff to study science and, unusually for this period, there was particular encouragement towards the study of the physical sciences. Elsie opted to study chemistry at Imperial College, graduating with a BSc in 1928 and a PhD three years later. Her sister studied mathematics for her BSc before taking an MSc in quantum mechanics and a PhD in nuclear physics. She achieved lasting fame under her married name of Eva Crane for her work on apiculture, while Elsie's interests developed from carbohydrate chemistry to the nutritional composition of foods.

Widdowson obtained her PhD for research, funded by the Department of Scientific and Industrial Research and carried out under the direction of Helen Archbold (later Porter, FRS 1956) in the Department of Plant Physiology at Imperial College. Here she used her chemical skills to develop a method for analyzing the carbohydrate content of apples. Her next post was at the Courtauld Institute at the Middlesex Hospital where she worked with Professor (later Sir) Edward Dodds on problems in human biochemistry. Despite her qualifications and publication record

European Women in Chemistry. Edited by Jan Apotheker and Livia Simon Sarkadi
Copyright © 2011 WILEY-VCH Verlag GmbH & Co. KGaA, Weinheim
ISBN 978-3-527-32956-4

Elsie May Widdowson
(https://www.imperial.ac.uk/publications/reporterarchive/0094/
news07.htm).

Widdowson struggled to find a job in 1933 and, on the advice of Professor Dodds, enrolled in the one-year dietetics diploma at King's College of Household and Social Science. It was while working in the main kitchen at King's College Hospital in preparation for this course that Widdowson first met Robert McCance, whose research caused him to bring joints of meat to the kitchen for cooking. She was able to draw on the expertise gained during her research on apples to correct some of his earlier work on the available carbohydrate of foods. McCance was impressed enough to seek funding from the Medical Research Council to employ Widdowson as his assistant and they embarked on further studies of food composition. Widdowson also completed her diploma in dietetics. It was her experiences during this course that inspired her to suggest to McCance the idea for a set of practical tables for the composition of British foods, which she believed would be more useful to dieticians than the American tables, that covered only raw foods, that were then currently in use. This project came to fruition in 1940 with the publication of the first edition of *The Chemical Composition of Foods*. Between 1934 and 1938 McCance and Widdowson continued to collaborate on a range of investigations on human diet and metabolism at King's College Hospital. In 1938 McCance was invited to become reader in medicine at the University of Cambridge and managed to persuade the Medical Research Council to maintain funding for his collaborations with Widdowson. At Cambridge they continued their research into human metabolism. This frequently involved a significant amount of self-experimentation that did not always go to plan.

With the outbreak of war the focus of research moved to an experimental study of rationing. As well as producing scientific results, this work led to the decision to fortify flour with calcium carbonate as a precaution against shortages of calcium in the diet in the event of a shortage of dairy products. After the end of the conflict,

McCance and Widdowson, now permanent members of the MRC staff, were asked to go to Germany to study the impact of undernutrition on the population. On her return to Cambridge in 1949 she resumed a project she had started before her departure, on the composition of the human body. In 1968 Widdowson moved to the Dunn Nutrition Laboratory as head of the Infant Nutrition Research Division. Her official retirement in 1973 meant a move to the Department of Investigative Medicine at Addenbrooke's Hospital in Cambridge where, for several years, she had laboratory space and PhD students. Later, when laboratory space was no longer available, she retained an office which she kept until her final retirement in 1988. She died in 2000 following a severe stroke.

Honors, slow in coming before her formal retirement, were showered on Widdowson afterwards. In 1975 the University of Manchester awarded her an Honorary DSc and in 1976 she was elected Fellow of the Royal Society. She was appointed Commander of the British Empire (CBE) in 1979 and Companion of Honour in 1993. Accolades from her fellow scientists included the James Spence Medal of the British Paediatric Association, First European Nutrition Award, Federation of European Nutrition Societies and the First Edna and Robert Langholz International Nutrition Award, American Dietetic Association Foundation. She served as President of the Nutrition Society (1977–1980), the Neonatal Society (1978–1981) and the British Nutrition Foundation (1986–1996). When the Medical Research Council established a new unit for Human Nutrition Research in Cambridge in 1998 it was named the Elsie Widdowson Laboratory. In 2000 the British government established a Food Standards Agency which named the library in its new building after her.

Throughout her career Widdowson's scientific work involved the use of detailed and well-planned experiments to provide evidence upon which practical interventions could be developed. Her initial research involved the use of her chemical expertise to analyze the carbohydrate chemistry of the apple during ripening and storage, part of a project intended to reduce the wastage of fruit by optimizing the conditions in which it was stored to minimize these changes. This led to her first publication in the *Biochemical Journal*. Her work at the Courtauld Institute at Middlesex Hospital produced a paper on biochemical aspects of nephritis. After she started working with McCance she combined work on food tables with investigations into the problem of salt deficiency in humans which contributed to an understanding of the importance of maintaining fluid and chemical balance, particularly in patients with diabetes. Their research later moved on to investigate the absorption and excretion of iron, using themselves as experimental subjects, and renal function, in particular the initially puzzling differences between children and adults. Widdowson also instigated dietary surveys that concentrated on individuals rather than families or households, as had previously been the norm.

After McCance and Widdowson moved to Cambridge they continued their research on absorption and excretion, turning their attention to strontium and maintaining the practice of self-experimentation. The outbreak of war caused them to direct their attention towards experimental studies of rationing and experiments on human endurance that were intended to provide data to determine the foods that best contributed to human efficiency but which made the most efficient use of the available shipping space. They also considered the issue of bread composition, and produced results that had a direct impact on food policy. After the end of the war Widdowson spent three years in Germany carrying out studies in orphanages on the relationship between diet and growth in children. Unexpected results during one of these studies led her to argue that environmental factors as well as nutritional ones were important for optimum growth, and that even well-fed children suffered from delayed growth if they were placed in a stressful environment.

Widdowson returned to Cambridge in 1949. Here she worked on the composition of the human body, particularly infants. This had a comparative perspective and extended to include research on energy intake and expenditure and the impact of litter size on the early development of a range of mammals. When she moved to the Dunn Nutrition Laboratory in 1968 she turned her attention to the composition of the adipose tissues of infants. This project was inspired by the observation that infant formulas in the UK and the Netherlands had very different chemical composition. Widdowson established that this resulted in significant differences in the composition of the body fat of infants in the two countries. These studies were extended to guinea pigs. She also investigated the impact of low birth weight on subsequent growth and development. After her formal retirement she collaborated with Olav Oftedal at Cornell University to investigate the suckling and growth of seals and black bears.

Elsie Widdowson was one of the most significant and productive figures in British nutrition research for over half a century. She contributed to over 600 publications alone and in collaboration with others, especially Robert McCance. These ranged across a variety of topics, from her earliest research on chemical aspects of apple physiology to her final work on the body composition of prenatal and early suckling animals. The apparent delay in achieving major honors such as election to the Fellowship of the Royal Society is perhaps explained by her long association with McCance and the difficulty of establishing their individual contributions. It has also been attributed to her self-effacing character and humility, characteristics that many women scientists of her generation adopted as a coping strategy in an often hostile employment climate in which self-promotion was discouraged in favor of quiet competency and a supporting role.

Literature

Ashwell, M. (2002) Elsie May Widdowson, CH, 21 October 1906–14 June 2000, *Biographical Memoirs of Fellows of the Royal Society*, **48**, 483–506

Ashwell, M. (ed.) (1993) *McCance and Widdowson: A Scientific Partnership of 60 Years, 1933–1993*, British Nutrition Foundation, London.

Whitehead, R. Widdowson, Elsie May (1906–2000) in *Oxford Dictionary of National Biography*, Oxford University Press, Sept 2004; online edn, May 2006, http://www.oxforddnb.com/view/article/74313 (accessed 30 July 2010).

Obituaries in The Times, Guardian, Independent, Daily Telegraph.

Bogusława Jeżowska-Trzebiatowska (1908–1991)

Henryk Kozlowski

■ Coordination and inorganic chemistry in Poland is closely associated with the name of Bogusława Jeżowska-Trzebiatowska, the professor of chemistry at the University of Wrocław. Professor Jeżowska-Trzebiatowska was a world famous scientist, one of the greatest personalities in Polish chemistry, an open-minded person, passionately devoted to science, the founder of the leading Polish School of Inorganic and Coordination Chemistry and the creator of modern fields including inorganic biological chemistry, biomedical chemistry and metal-based catalysis.

She promoted 71 PhD students, 34 of whom later became professors. At present her school consists of 70 professors engaged in different fields of chemistry. She published around 600 scientific papers, 33 monographic books and reviews. One of the most characteristic features of her School was interdisciplinary research involving chemistry, physics, biochemistry, biology and medicine as well as the technical sciences.

Professor Jeżowska-Trzebiatowska was born in Stanislavov near Lvov (now Ukraine) on November 19, 1908. As a young girl, she was fascinated by the humanities but while in school she began a second love-affair with chemistry and physics. A substantial impact on her passion for chemistry and physics derived from the personality and successes of Marie Skłodowska-Curie, who visited Lvov and gave a lecture in the City Hall. This encouraged the young lady to deide to study chemistry at Lvov Technical University against her parents' wishes. During the academic year 1926/1927 Bogusława Jeżowska, along with nine other girls, became a newcomer among 100 students in the Faculty of Chemistry.

While a third year student she met Professor Jakób who offered her the position of an assistant. As soon as 1931, she published her first scientific paper "Six-valent molybdenum complexes with hydroxylamine". However, the real love of young Mme Jeżowska was rhenium, which was brought to Lvov in 1931 by Professor Jakób. In 1932 she published her first paper "Über das fünfwertige Rhenium". As a young researcher she published a series of papers on the physico-chemistry of rhenium complexes, especially concerning electrochemical and chemical mechanisms of the reduction of per-rhenates to rhenium +5. The publications were very

European Women in Chemistry. Edited by Jan Apotheker and Livia Simon Sarkadi
Copyright © 2011 WILEY-VCH Verlag GmbH & Co. KGaA, Weinheim
ISBN 978-3-527-32956-4

Bogusława Jeżowska-Trzebiatowska
(Photo from author's private collection)

well accepted, to such an extent that during the lectures given at the Paris Sorbonne University she got the nickname "La mère du rhenium". Her PhD thesis also involved this topic. She was the first woman who defended a PhD thesis in the Technical University of Lvov in 1935. In her memoirs she recalls "The defense of the doctoral thesis was described in the newspapers and the main University Hall was full of various observers. The young candidate for doctor was dressed in an elegant black dress with a red rose attached to it".

In 1935 Bogusława Jeżowska married Włodzimierz Trzebiatowski, a distinguished specialist in solid-state physical chemistry and this marriage radically affected her approach to chemistry. Unfortunately, in 1939, the Soviets entered Poland and Lvov; the Second World War began and this stopped research for a rather long period of time. During the war, she first worked in a cake shop and then, being afraid of the spectre of transfer to Germany, she started to work in the German Consortium of Lvov Factories where Poles were employed. She became a director of the chemical factory Hohere Alcohole. In 1942 Dr. Jeżowska-Trzebiatowska started her cooperation with AK, the dominant Polish resistance movement in World War II in German-occupied Poland. Being a person who had easy access to chemicals, her job was the production of explosives. For her work in the resistance organization, she was rewarded with one of the highest distinctions of the Polish Underground State.

In 1991, the Institute of National Remembrance in Jerusalem – Yad Vashem – awarded Bogusława Jeżowska-Trzebiatowska with the 'Righteous Among the Nations' medal for saving the life of Dr Emil Taszner, later Professor of Chemistry at the Technical University of Gdansk, the creator of the Polish school of peptide chemistry, who was hiding in the factory from December 1942 to August 1944. He also hid in Dr. Trzebiatowska's apartment.

After the re-occupation of Lvov by the Russians and the transformation of the German factory to "Chemtrud", she worked there for some time, and, after the war, in December 1945 she moved to Wroclaw together with her husband. The conditions were very difficult; at that time, 80% of the city of Wroclaw did not exist.

Along with many other professors from Lvov she was a pioneer of the Polish science revival, including the organization of academic life from scratch. Initially, she was the organizer of scientific research and teaching at the combined universities, the University of Wroclaw and the Wroclaw University of Technology (she headed the Faculty of Inorganic and Analytical Chemistry at the Department of Pharmacy, the Chair of General Chemistry and the Chair of Inorganic Chemistry at Wroclaw University of Technology, later the Chair of Chemistry of Rare Elements). In 1951, she organized a new faculty – the Faculty of Chemistry – at the Department of Mathematics–Physics–Chemistry at the University of Wroclaw. In the years 1958 to 1962 she held the position of dean of this faculty. The three Chairs were established there – Inorganic Chemistry (headed by Professor Trzebiatowska), Organic Chemistry and Physical Chemistry. In 1969 from the merger of the three Departments of Chemistry, with great organizational effort, she created the Institute of Wroclaw University – her pride and glory. She was anxious to develop new research methods and in this way to locate "her Institute" at the forefront of the worlds scientific institutions. In the years 1969 to 1979 Professor Jeżowska-Trzebiatowska held the position of head of the Institute of Chemistry and of the Department of Inorganic Chemistry at the University of Wroclaw.

Together with organizing activity, she worked intensively in the scientific field. Systematically, and with stubbornness, she aimed higher and higher in her scientific career; she made a habilitation in 1949 based on studies on the chemistry and physical chemistry of rhenium. In 1954, she was given the title of Professor. She was involved in international lecturing at universities in Paris, Rome, Florence, Geneva, Berkeley, Ann Arbor, Urbana-Champaign, Los Angeles, Tokyo, Toronto, Melbourne, Toulouse, Budapest, Prague, Athens, Nanjing, Porto, London, Stockholm, Vienna, Zurich, Moscow, Berlin, Dresden, Halle, Leipzig, Leningrad and so on. She organized numerous international and national scientific conferences, she invited scientists from the world's leading research centers to her Institute.

The organization of scientific meetings, and the integration of the scientific environment was one of Professor Trzebiatowska's major objectives. Her most important aim, which she fully achieved, was the introduction of the Wroclaw and Polish chemistry into world science. She developed an exceptionally wide cooperation with many research centers in the world, which resulted in departures of the Institute's employees for internships, numerous conferences and symposia.

She was strongly associated with the Polish Academy of Sciences (PAN). From 1967, she was a member of the Academy, and from 1978 until her death she was the head of the Wroclaw Branch of PAN. From 1967, she was also the head of the Department of Structural Chemistry, Institute of Low Temperatures of PAN in Wroclaw.

She was a member of many committees and scientific societies including: the International Union of Pure and Applied Chemistry (IUPAC), the Academy of Leopoldinum, the Council of Higher Education, the Science Committee of Chemical Sciences of PAN, the European Society of Physics, the Committee of Physics, Biochemistry and Biophysics, and head of the PAN Spectroscopy Committee.

Her efforts and work for science were widely noticed and appreciated. She was the winner of many National Awards, Awards of the State Council on the Peaceful

Use of Nuclear Energy, the Special Prize of Polish Science and the prizes awarded by the Ministry of Higher Education. She was also awarded honorary doctorates at: the Technical University in Bratislava (1971), the Lomonosov Moscow State University (1979), the Wroclaw University of Technology (1980) and the University of Wroclaw (1981) and honored with many medals.

The spectrum of scientific interests and the research undertaken by Professor Bogusława Jeżowska-Trzebiatowska were impressively broad. The chemistry of rhenium was not her one and only love and passion. Indescribably scientifically active, she undertook various other novel and up-to-date research problems. These were often unique and pioneering studies in various areas of chemistry and physical chemistry, and particularly in coordination chemistry. She led and initiated so many research projects; among others, were studies on the anti-ferromagnetism of complex compounds, the magnetism and spectroscopy of f-electron elements, the oxygen bond (as the creator of the oxygen bridge theory, she became a prominent figure in the history of world science), and the so-called hydrogen bond.

She devoted many years of work to the electronic and molecular structure of complex compounds, studying the coordination properties of d-electron elements, doing research in the field of bioinorganic chemistry and biophysics, and radiation chemistry, and studies on the activation of small gas molecules or catalytic processes. With her initiative and her participation, various spectroscopic methods (X-ray structural analysis of complex compounds, or the radioisotopic studies of the structure and mechanisms of chemical reactions) evolved in the Institute of Chemistry. She also initiated intense studies on the spectroscopy and luminescence of coordination compounds. Among other things, several new laser-active materials based on lanthanide compounds were discovered. She was able to gather an energetic team of scientists around her, people sharing her creative passion, who undertook and continued those studies. She had the gift to inspire people with a craving for knowledge.

The educational achievements of the Professor are not only excellent lectures and the promotion of over 70 PhDs but also the patronage of numerous high schools in our region. It is impossible not to mention the cooperation with industry, in which Professor Trzebiatowska took part, and various studies and elaborations made especially for the copper industry.

The Professor was extremely devoted to her work, her students and the Institute. She demanded a lot from others, but even more from herself. She was pleased with success of her coworkers and encouraged their self-development. She used to say "What really matters is simply to be able to work, create something, serve not only science and discoveries, but to serve people, to serve my students – this is the greatest satisfaction. What saddens is the shortness of time which was given to man".

She was a beautiful and elegant woman of incredible class and perseverance. High psychological resistance, courage, strong character, perfection, the ability to deal with any fate, and the constant willingness to help others added up to her being an amazing person.

Professor Bogusława Jeżowska-Trzebiatowska died a tragic death on 16 December 1991. The foundations of chemistry created by her serve and will serve many generations of Polish scientists. With the death of Professor Bogusława Jeżowska-Trzebiatowska and her husband Professor Włodzimierz Trzebiatowski, not only Wroclaw science, but Polish and world science suffered a great loss. Such titans of work and knowledge are rarely born.

Literature

Kozłowski, H. and Legendziewicz, J. (1993) *Nauka Polska*, **2–3**, 201–205.

Stasicka, Z. and Ziółkowski, J. (2005) *Coord. Chem. Rev.*, **249**, 2133–2143.

Ziółkowski, J. (2000) *Coord. Chem. Rev.*, **209**, 15–33.

Yvette Cauchois (1908–1999)

Christiane Bonnelle

■ The international renown of Yvette Cauchois originated in her first research work. During the preparation of her doctoral thesis (1933), she made a high-resolution X-ray spectrograph with high luminosity, known as the "Cauchois spectrograph". This apparatus is still the best performing instrument for the high-energy X-ray and gamma ranges. Thanks to the efficiency of this spectrograph, she was able to measure for the first time numerous X-ray emission lines of very low intensity belonging to heavy and rare elements. She contributed greatly to the development of X-ray spectroscopy and the understanding of electronic energy level structures.

From 1953, as the director of the Laboratory of Physical Chemistry (LPC) at the Sorbonne, and later at the University Pierre et Marie Curie, both in Paris, she initiated numerous research programs, among which were the development of soft-X-ray and UV spectroscopies for the study of light elements and chemical binding in solids, X-ray microscopy, electron–matter interactions and the utilization of synchrotron radiation as the light source in a wide energy range from X-rays up to UV. These studies have made possible fundamental advances in various areas of physical chemistry.

Yvette Cauchois was born in December 1908 in Paris, where she lived all her life. She obtained her first degree in Physical Sciences at the Sorbonne in June 1928 and in July she was accepted in Professor Jean Perrin's laboratory. She was 19 years old but she had been attracted to science even as a child. Under Francis Perrin's guidance she started research on fluorescence and obtained a Diplôme d'études supérieures in 1930. She then turned to X-ray spectroscopy and defended her thesis, entitled "An extension of X-ray spectroscopy: a spectrometer focused with a curved crystal; X-ray emission spectra from gases" in July 1933. She was 24 years old.Very rapidly, her work brought her international attention.

The first X-ray spectrographs used Bragg reflection on flat crystalline slates. The resolving power was controlled by a slit and a good resolution was obtained only at the expense of the luminosity. Among numerous attempts made to improve the quality of these instruments, the suggestion by H.H. Johann to use the reflection from a concave surface of a curved crystalline slate caught Cauchois's attention. In

European Women in Chemistry. Edited by Jan Apotheker and Livia Simon Sarkadi
Copyright © 2011 WILEY-VCH Verlag GmbH & Co. KGaA, Weinheim
ISBN 978-3-527-32956-4

Yvette Cauchois (illustration provided by author)

this geometry, it was possible to concentrate the reflected radiation from a large X-ray beam. But the resolution was very poor in the hard X-ray range because of the smallness of the Bragg angles. Thus this method was only useful in the soft X-ray range.

Cauchois's idea was then to use the reflection from crystalline planes oriented either perpendicularly, or obliquely, with respect to the surface of the curved crystalline slate. The rays strike the convex face of the crystal; the reflected radiation leaves the concave face in the geometry of transmission and impinges on the detector in the region corresponding to large Bragg angles. The reflected radiation converges on a narrow zone, thereby increasing the luminosity. The apparatus had both high resolving power and a large gain in luminosity with respect to other apparatuses. Yvette Cauchois formulated the principle of this apparatus [1], and demonstrated its interest for high-resolution spectroscopy and also as a monochromator for the study of X-ray diffraction [2]. She used the curved crystal technique to create the first focusing systems for forming X-ray images of emitting or opaque real objects [3].

Thanks to the high quality of this spectrograph, she was able to observe the X-ray spectra emitted from heavy rare gases for the first time. These spectra were unknown until this date because they required an instrument capable of measuring accurately the emission of X-ray sources of very low intensity. Her observations were of immediate interest to the scientific community and the use of this spectrograph extended rapidly to all the research centers involved in X-ray spectroscopy and its application, for example in the laboratories of Professor Mane Siegbahn (Upsala), Professor Kramers (Leyde), Professor Zeeman (Amsterdam), and in numerous other European countries, in the USSR, Japan, the USA, and Australia.

Due to this gain in luminosity, Cauchois measured with high precision the entire emission K spectra of argon and xenon, including the lines of very low intensity. She then studied systematically weak lines, called satellites, accompanying the X-ray normal lines and corresponding to transitions in multiply ionized atoms. Most of these lines were then unknown. She observed a new family of satellite lines and determined for the first time the core level energies for multiply ionized heavy atoms.

In collaboration with H.H. Hulubei, Yvette Cauchois then undertook the detection of rare elements. First, they proved the presence of the element of atomic number 93 (Np) in uranium minerals. They observed the polonium L spectrum from samples weighing only a few micrograms, establishing many lines of this element and confirming that it has indeed the atomic number 84. Then they analyzed the natural emissions of the daughter elements of radon. They identified the already known elements 82, 83, 84. Moreover, they also discovered the unknown element 85 from observation of its three principal lines whose wavelengths were measured with a high precision because of their sharpness [4]. It is remarkable that the wavelengths measured for these three lines attributed by Hulubei and Cauchois to the element 85 are in perfect agreement with up to date calculations using a multiconfiguration Dirac–Fock (MCDF) program including Breit interaction and QED corrections, that was not available then, showing that there could have been no doubt that these lines originated from the element 85. A debate has been opened recently by Thornton and Burdette [4] about these old observations from which it

Plutonium L emission spectrum: dipole transitions, some quadrupole ones, some satellites and transitions from valence electrons are observed for the first time. The sample is a PuO_2 screen; it is excited by fluorescence with the help of an X-ray tube. The reflecting crystal is a mica slate parallel to 100 planes. The spectral range is 0.5–1 Å. The oblique lines are due to reflections on other crystalline planes [5].

becomes apparent that Hulubei and Cauchois were the first to have proved the existence of the element 85.

We cannot cite the many different research interests that occupied her during her long scientific career. Among the most significant works, we mention studies of the chemical bond from the X-ray absorption spectra [6], the first observation of the X-ray reflection from crystals in their anomaly region [7] and an excellent analysis on the electron interactions with matter including a chapter on synchrotron radiation [8].

In 1932, she obtained one of the first research scholarships from the new French national centre of sciences (Caisse Nationale des Sciences) founded by Perrin in 1930, and consequently became a permanent member of this centre (1937) and then in its successor the National Centre for the Scientific Research (CNRS), founded in 1939. As a young researcher at the Laboratory of Physical Chemistry (LPC), headed by Jean Perrin, she made many contacts with the illustrious foreign scientists who visited the laboratory. The "Monday teas" were the meeting place for important personalities of the Parisian scientific and cultural society of that period. She always spoke with enthusiasm of the pre-war years that brought together Jean Perrin, Marie Curie, Irène Joliot-Curie and Frédéric Joliot, Paul Langevin, and young research workers like Francis Perrin, Pierre Auger, Louis Leprince-Ringuet and many others.

During the war years, 1940–1945, when Jean Perrin had to leave for the United States, she was responsible for the continuity of the research at the LPC. She remained there when she became an assistant professor at the Sorbonne in 1945 and full professor in 1951. In 1953 she became the Director of the laboratory and she was named in the Chair of Physical Chemistry. The building of the LPC became too small for the group of Parisian physical chemists that she succeeded in attracting around her. Thus, in 1960, she founded the Physical Chemistry Centre (Centre de Chimie Physique) at Orsay, near Paris, and she directed both for ten years. She was President of the French Society of Physical Chemistry from 1975 to 1978. Retired in 1978, she continued to work at the laboratory until 1990, she so cherished this place to which she consecrated the best part of her life. From 1990, she became bedridden due to severe arthritis. At the end of August 1999, she went on a trip to Romania and died there, at 90 years old, on November 19, 1999.

Yvette Cauchois was an Officer of the Legion of Honour, Commander in the *Ordre des Palmes académiques* (French decoration for services to education), and Officer in the National Order of Merit. She was nominated Doctor *honoris causa* of the University of Bucharest in 1993. Her research work was rewarded by many prizes: prize of the Société Française de Physique (1933), three prizes of the *Académie des sciences* (1935, 1936, 1946), the Medal of the Czechoslovak Society of Spectroscopy (1974), the Gold Medal of the University of Paris (1987).

Literature

1. Cauchois, Y. (1932) Spectrographie des rayons X par transmission d'un faisceau non canalisé à travers un crystal courbé (1). *J. Phys., série VII*, **III**, 320; Cauchois, Y. (1933) Spectrographie des rayons X par transmission d'un faisceau non canalisé à travers un crystal courbé (2). *J. Phys., série VII*, **IV**, 61.

2. Cauchois, Y. (1932) Une nouvelle méthode d'analyse des poudres cristallines par les rayons X, utilisant un monochromateur à crystal courbé. *Compt. Rend. Acad. Sci.*, **195**, 228.

3. Cauchois, Y. (1950) Sur la formation d'images avec les rayons X. *Rev. Opt.*, **29** (3) 151.

4. Thornton, B.F. and Burdette, S.C. (2010) Finding eka-ionine: discovery priority in modern times", *Bull. His. Chem.*, **35** (2), D76.

5. Cauchois, Y. and Manescu, I. (1956) Spectres de fluorescence L du plutonium, *Compt. Rend. Acad. Sci.* **242**, 1433

6. Cauchois, Y. (1954) Spectres X et liaison chimique, *J. Chim. Phys.*, **51**, D76.

7. Cauchois, Y. (1956) Distribution spectrale dans les régions d'absorption propre de divers cristaux, *Compt. Rend. Ac. Sc.* **242**, 100.

8. Cauchois, Y. and Heno, Y. (1964) *Introduction à l'Emploi de Rayonnements en Chimie Physique. Cheminement des Particules Chargées*, Gauthier-Villars, Paris.

Marguerite Catherine Perey (1909–1975)

Jean-Pierre Adloff

■ Two years after Henri Becquerel's discovery of radioactivity in 1896, Pierre (1859–1906) and Marie Curie (1867–1934) found two new elements, polonium and radium, solely by virtue of their emission of invisible rays. This was followed by the discovery of three more radioactive elements: actinium, radon and protactinium. Together with uranium and thorium, also radioactive but known earlier, these elements possessed a common property: they occupied vacant positions beyond bismuth in the terminal part of the periodic table. Hence, it could be predicted that *all* elements beyond bismuth should be radioactive. In 1939, Marguerite Perey discovered the element 87, which was one of the missing elements in the periodic chart. Dmitrii Ivanovich Mendeleev (1834–1907) had predicted that this element named 'ekacaesium' would be the most electropositive of all elements. This element was named Francium.

Marguerite Catherine Perey, the youngest of five children, was born on October 19, 1909 in Villemomble, near Paris. A stock market crash and the death, in March, 1914, of her father, the proprietor of a flour mill, created financial difficulties for this middle-class protestant family, which prevented the children from pursuing any higher education. Perey attended a technical school for girls (*École d'Enseignement Technique Féminine*), a state-recognized school where she graduated as a chemist in 1929. The same year, she was hired by the Institut du Radium in Paris, where her intelligence, skill and eagerness to learn and understand brought her to the attention of the Director, the 1903 Nobel Physics and 1911 Nobel Chemistry laureate Marie Skłodowska-Curie. Perey soon became her personal assistant and confidant. Her first years spent with Marie Curie might be regarded as a premonitory sign of destiny and the first step toward a major discovery.

After the discovery of francium, Perey undertook university studies at the Sorbonne during World War II. The same year she defended her doctoral thesis (1946), she was appointed *Maître de Recherches* at the national centre of scientific research (CNRS). In 1949 she was called to a new chair of nuclear chemistry at the University of Strasbourg, at the time the only such chair in France outside Paris. In 1957 she became Director of the Department of Nuclear Chemistry at the newly found-

European Women in Chemistry. Edited by Jan Apotheker and Livia Simon Sarkadi
Copyright © 2011 WILEY-VCH Verlag GmbH & Co. KGaA, Weinheim
ISBN 978-3-527-32956-4

Marguerite Catherine Perey, illustration provided by the author.

ed Nuclear Research Centre. She never married but devoted her time to her scientific and educational responsibilities to national (CNRS) and international (IUPAC) Committees. Knight (1958) then Officer (1960) of the Legion of Honour, Perey received numerous honors and awards, including the Grand Prix de la Ville de Paris (1960) and two prizes from the French Academy of Sciences (1950 and 1960). In 1962, she was elected corresponding member of this academy, which had been closed to women (even to Nobel laureates Marie Curie and Irène Joliot-Curie) since its founding in 1666.

Shortly after 1946, Perey noticed a burn developing on her left hand, which was diagnosed as cancer caused by her many years of working with radioactive elements, in particular actinium. After several long stays in hospital she moved to Nice but maintained close contact with her laboratory. Long periods of illness alternated with short weeks of remission as the disease, which had claimed the lives of Marie Curie and her daughter Irène Joliot-Curie, progressed. In 1967 Perey attended the centenary celebration of Marie Curie's birth at Warsaw, her last appearance with the international community of nuclear scientists. By July 1973 her disease had become more acute, forcing her to stay in the Curie Hospital in Paris and finally in the Clinique du Val de Seine at Louveciennes, where she died on May 13, 1975, at 65 years old, one of the last survivors of the pre-World War II radiochemical pioneers from the *Laboratoire Curie*. Her friend and 1966 Nobel Physics laureate Alfred Kastler (1902–1984) read her obituary before the Académie des Sciences. He had presented a last distinction as Commander in the national order of merit to her at her bedside precisely one year earlier.

When Perey began to work at the *Institut du Radium* in 1929, the first task assigned to her was the purification of actinium (^{227}Ac), a radioactive element discovered by André Debierne (1874–1949) in 1899. Actinium is always mixed with rare earths (lanthanides), from which it is very difficult to separate. The actinide series had not been studied to the same extent as the other two naturally radioactive families, *viz.* the radium and thorium series. Even the half-life of actinium was uncertain. The element is much "rarer" than the accompanying rare earths (lanthanides). Perey had to concentrate actinium among the lighter lanthanides, an operation requiring hundreds of fractional crystallizations. Actinium-specific radiation, a very soft β-ray, was not detectable. Progression in the concentration of the element was monitored from the β- and γ-rays emitted by its radioactive descendants, and a waiting time of three months was required to permit establishment of radioactive equilibrium. Conscientiousness, perseverance and enthusiasm were prerequisites for such a demanding task.

By the mid-1930s Perey succeeded in preparing the most intense source of actinium ever available. Marie Curie had asked for this sample to measure the emission spectrum of the element, the test required for its definitive identification. Perey, who meanwhile had acquired good knowledge of spark spectroscopy, actively participated in the project, which was interrupted by Marie Curie's death on July 4, 1934. The loss of her 'patron' was a severe shock for Perey, and on many occasions she recalled with considerable emotion the period of five years that she had spent in close and nearly daily contact with Marie Curie.

Perey now began to work under the direction of Debierne and Marie Curie's daughter, Irène Joliot-Curie (1897–1956). Both were interested in actinium and, independently of and unknown to each other, asked Perey to pursue the concentration and purification of actinium. Irène Joliot-Curie wanted to determine the precise half-life of actinium, whereas Debierne was involved in the search for 'neo-radioelements' which did not exist. During the autumn of 1938, Perey observed that actinium, freshly purified from all radioactive daughters, emitted a hitherto unknown β-radiation, which increased in intensity over two hours and then remained constant. During the following hours and days the β-activity increased again as the long-lived daughters were slowly formed. Perey's thoroughness and speed in performing the experiments enabled her to observe the phenomenon, which had remained undetected for 40 years by earlier and less skilful radiochemists.

In January, 1939, after numerous tests, Perey concluded that part of the decay process of ^{227}Ac leads to a β-emitting radioelement. This radioelement exhibited chemical properties of an alkali metal element, which conceivably could only be the long-sought 'eka-caesium' with atomic number 87. Shortly afterward, Perey unequivocally established its genesis through emission of α-rays by ^{227}Ac. Starting from atomic number 89, this α-decay led to the still vacant position 87 in the periodic table. As a modest 29-year-old technician without a university degree, Perey had discovered the first isotope of 'eka-caesium' with mass number 223. Following the tradition of the time, she named it actinium K (AcK). Her careful measurements indicated that 1.2% of the actinium atoms decay to eka-caesium, with a half-life of 21 min, close to the most recent values (22 min and 1.38%). The discovery of element 87, "Sur un élément 87, dérivé de l'actinium", was announced cautiously on January 9, 1939 at the weekly session of the French academy of science by the 1926 Nobel Physics laureate Jean Perrin (1870–1942).

After the discovery of AcK, Debierne and Irène Joliot-Curie encouraged Perey to undertake university studies while pursuing her experiments. She received her *licence* diploma and was now qualified to defend a thesis titled "*L'élément 87: Actinium K*," on March 21, 1946. The last line of the thesis reads thus: "The name Francium Fa, is proposed for box 87". This name was officially adopted a few years later, but the symbol was changed to Fr, and AcK was changed to ^{223}Fr. The members of the examination board included Debierne and Irène Joliot-Curie. What Perey appreciated most after the dissertation was Irène's comment: "Today my mother would have been happy".

Francium is the fourth natural radioelement discovered in France after polonium, radium, and actinium and is the last element to be discovered in nature. It is the rarest and most unstable of all naturally occurring elements. Its entire content in the Earth's crust at any time is several hundred grams, compared to 7400 tons for the Curie's polonium. The most recent nuclide chart lists 24 isotopes of francium, among which Perey's AcK (half-life 22 min) is the longest lived. All elements beyond francium (87), up to dubnium (105), have isotopes longer lived than AcK.

When the Chair of Nuclear Chemistry at Strasbourg was offered to Perey, she accepted the nomination in a spirit of fidelity "trying to communicate to a new team the eagerness of rigorous and joyful work and in this way to do homage to Marie Curie, my beloved and venerable Master". She was now interested in the biological applications of francium, hoping that it would be useful for the establishment of an early diagnosis for cancer. Despite encouraging results, the project was abandoned because of the lack of a sufficient amount of actinium and of the little interest shown by physicians.

Perey benefited greatly from Marie Curie's prestige. She inspired respect and admiration from students, collaborators, and colleagues. However, the two women possessed little in common. Perey's initial scientific background was elementary, while Curie had university degrees in mathematics and physics, and her knowledge encompassed the most recent theories and findings of her time. Curie's discoveries of polonium and radium resulted from reasoning on previous observations, while the discovery of francium is a perfect example of serendipity, that is the accidental discovery of things not sought. Both scientists suffered from radiation sickness and died at nearly the same age, but Curie worked until the last weeks of her life, while Perey struggled for 16 years against the disease.

Acknowledgement

George B. Kauffman, Professor of Chemistry Emeritus at California State University, Fresno and a Guggenheim Fellow, is thanked for comments and revision of the text.

Literature

Adloff, J.P. and Kaufmann, G.B. (2005) Marguerite Catherine Perey (1909–1975). in *Out of the Shadows: Contributions of 20th Century Women to Physics* (eds N. Byers and G. Williams); Cambridge University Press, Cambridge, England, pp. 371–384.

Adloff, J.P. and Kauffman, G.B. (2005) Francium (Atomic number 87), The Last Discovered Natural Element. *Chem. Educ.*, **10**, 387–394.

Adloff, J.P. and Kauffman, G.B. (2005) Marguerite Perey (1909–1975): A Personal Retrospective Tribute on the 30th Anniversary of Her Death. *Chem. Educ.*, **10**, 378–386.

Adloff, J.P. and Kauffman, G.B. (2005) Triumph over Prejudice: The Election of Radiochemist Marguerite Perey (1909–1975) to the French Académie des Sciences. *Chem. Educ.*, **10**, 395–399.

Kastler, A. (1975) Notice nécrologique sur Marguerite Perey (1909–1975). *Compt. Rend. Ac. Sc.*, 280, vol. Vie académique, 124–128.

Kaufmann, G.B. and Adloff, J.P. (1993) Marguerite Catherine Perey (1909–1975) in *Women in Chemistry and Physics* (eds L.S. Grinstein, R.K. Rose, and M.H. Rafailovich); Greenwood Press, Wesport, CT, pp. 470–475.

Perey, M. (1946) L'élément 87: Actinium K. *Thesis*, Faculté des sciences de l'Université de Paris, March 21, 1946. *J. Chim. Phys.*, **43**, 152–168.

Perey, M. (1939) Sur un élément 87, dérivé de l'actinium. *Compt. Rend. Ac. Sc.*, **208**, 97–99.

Filomena Nitti Bovet (1909–1994)

Marco Ciardi and Miriam Focaccia

Filomena Nitti was a scientist at the forefront of the development of pharmacology and therapeutic chemistry after the Second World War. Together with her brother, Federico, and husband, Daniel Bovet, she conducted much important research in fields ranging from general pharmacology to sulfonamide chemotherapy, pharmacology of the vegetative nervous system, anti-allergy treatment, the use of synthetic curares in anesthesiology, hormone balance modifiers and pharmacology of the central nervous system. Her husband was awarded a Nobel Prize in 1957. Many colleagues felt that Filomena made a major contribution to this achievement.

Filomena was born on 10 January, 1909, the daughter of Francesco Saverio Nitti – Prime Minister in 1919 and 1920 and a renowned economist – and Antonia Persico. She had one sister, Maria Luigia, and three brothers: Vincenzo, Giuseppe and Federico, a doctor, with whom she shared much of her scientific career.

She spent her childhood between Naples, where she lived with her paternal grandparents, and Rome, where her parents had their main residence. She was reunited with the latter most often during long summer holidays spent in their house in Acquafredda.

Filomena and Daniel Bovet (http://www.pictokon.net/
bilder/2007-06-g/bovet-daniel-und-filomena-bovet-nitti.html).

European Women in Chemistry. Edited by Jan Apotheker and Livia Simon Sarkadi
Copyright © 2011 WILEY-VCH Verlag GmbH & Co. KGaA, Weinheim
ISBN 978-3-527-32956-4

Thirteen year old Filomena's life changed dramatically after 1922, following events connected to the coming of Fascism. The Nittis were subjected to frequent attacks by fascist squads. Their house in Rome was attacked, plundered and destroyed. They consequently went back to Naples, but everyday life did not turn out to be any easier. Even going to school was a difficult undertaking: Filomena and Federico became the targets of a series of attacks. In a climate like this, where Francesco Saverio Nitti was often forced to go into hiding, the decision was taken to leave Italy. He left with his family for Zurich and then they moved to Paris.

Filomena enrolled at an Alliance Française evening school to learn French well. She then managed to get a place in the Sevigné, a "well-known" secondary school. After leaving school she enrolled in the Faculty of Natural Science. It was around the same time that her political militancy started. In around 1930 she entered the youth section of the communist party. She then left for Russia, where she worked at both the "Journal de Moscou" and Red Aid, at the time managed by Elena Stassova.

On returning to France she worked as a chemical analyst for a couple of years, before joining the Pasteur Institute (1938), first "as a guest" and then with a grant.

The Pasteur Institute had a wonderful reptile house, where the young researcher enjoyed carrying out valuable studies for her thesis on cobra poison. She was fascinated by the creatures she was studying and refused to have bags, shoes or clothes made from reptile leather. Her brother had already been working at the Institute for a number of years in the Laboratory of Therapeutic Chemistry alongside Daniel Bovet. The young researcher's meeting with the latter changed her destiny and the two got married in 1939. They shared a life together distinguished by scientific excellence and a passion for research.

The couple moved to Italy in 1946, when Domenico Marotta invited Daniel Bovet to become Head of the Laboratory of Therapeutic Chemistry at the ISS (Italian National Health Institute).

After he resigned from his post at the ISS, Filomena Nitti joined the Italian National Research Council in 1964 and stayed there until 1975.

Filomena Nitti was one of the leading players in the golden age of pharmacology and the development of therapeutic chemistry. She was introduced to this branch of research as a result of her PhD studies: the hemolytic action of cobra poison was the starting point that led her to conduct in-depth studies into the way in which other toxins act on the body, with a view to potential treatments.

The trio of Federico, Filomena and Daniel were a close-knit enterprising team. In the difficult period of German occupation they used the only instrument left available to them – a radio – to keep up to date on research developments in Great Britain and the United States. They dedicated themselves to growing strains of penicillin and managed to produce small quantities of them using purely handmade equipment to supply the French liberation forces.

In Paris, Nitti was a leading player in the construction of a new path in experimental medicine. These were the years in which patient laboratory research conducted in conditions that were frequently difficult laid the foundations for the acquisition of knowledge, later summarized in a book written jointly by Filomena and Daniel and published in 1948 *Structure et Activité Pharmacodynamique des Médicaments du Système Nerveux Végétatif.* This work served as a launch pad for the evolution of research work in the following decades, both in France and internationally. The book earned a reputation as a "Bible" within the scientific community.

The Nitti-Bovets' arrival in Italy marked an important moment in the post-war process of revival in the country. Their laboratory became a center of excellence for Italian pharmacological research. It became a destination for scholars from all over the world, including Nobel prize winner Boris Chain.

Filomena played a crucial role in the Rome-based institute: she welcomed newcomers, weaned them and got them started on research work. She was responsible for running the so-called "elementary school" at the pharmacological laboratory, supervising the apprenticeships of young talents. In her role as a "primary school teacher" she trained other leading female protagonists in the life of the ISS in that period: Maria Marotta, Maria Ada Iorio, Wanda Scognamiglio, Marisa Bizzarri, and Maria Amalia Ciasca.

She played a vital role in therapeutic chemical research in the first half of the twentieth century. The dedication she showed from her years at the Pasteur Institute onwards was also a determining factor in the awarding of the Nobel Prize to her husband, Daniel Bovet, in 1957. It was not by chance that the Italian psychiatrist Ugo Cerletti wrote a congratulatory letter to the married couple in the same year: he addressed both of them and applauded the fact that the awarding of the Prize would help finance a research enterprise closely shared and *built jointly.*

Literature

Bignami, G. (1993) Ricordo di Daniel Bovet, in *Annali dell'Istituto Superiore di Sanità*, **29**, suppl. n.1.

Bignami, G. and Carpi De Resmini, A. (2005) *I Laboratori di Chimica Terapeutica dell'Istituto Superiore di Sanità*, Istituto Superiore di Sanità, Roma.

Gobetti, C. (1986) Conversazione con Filomena Nitti, *Mezzosecolo. Materiali di Ricerca Storica*, pp. 397–430.

Passione, R. and Bovet, F.N. (2008) in *Scienza a Due Voci. Le Donne Nella Scienza Italiana dal Settecento a Novecento*, (eds V. Babini and R. Simili) (www.scienzaa2voci.unibo.it).

Bianka Tchoubar (1910–1990)

Didier Astruc

■ Bianka Tchoubar, of Karait origin (a Jewish sect from Babylon), born in Ukraine, arrived, at the age of 14, in Paris where she pursued her career in chemistry. Of legendary strength of character, passion, generosity and scientific rigor, she has been an emblematic and leading scientific personality in the French chemistry of the twentieth century. She played a major role as a scientific director, stimulating research by her passion, abnegation and scrupulous exactingness. She was outstandingly creative since the 1930s in bringing new ideas in organic reaction mechanisms, for which she had to fight continuously until the 1950s against the conservative ideas of several illustrious French organic chemists. She was a very charismatic Director of a very large research CNRS center at Thiais between 1968 and 1978. Her book on *Reaction Mechanism in Organic Chemistry* (1960) is very famous and has been translated into six languages.

Bianka Tchoubar was a person of exceptional human and scientific dimension. As a chemist, she was the pioneer of modern ideas in French organic chemistry in the middle of the twentieth century. She pioneered rational views of organic reaction mechanisms, in particular the crucial roles of ions as reaction intermediates. She was especially influential in the French community, because she was the very charismatic Director of the largest French organic chemistry Institute in Thiais, a Parisian suburb, between 1968 and 1978.

Bianka Tchoubar (illustration provided by author).

European Women in Chemistry. Edited by Jan Apotheker and Livia Simon Sarkadi
Copyright © 2011 WILEY-VCH Verlag GmbH & Co. KGaA, Weinheim
ISBN 978-3-527-32956-4

Paying very little attention to her own interests, Bianka Tchoubar had interactions with others whatever their social background. Of immense intelligence, intuition and culture, she had the ability to develop the deepest reasoning with scrupulous acuity and exactingness. She had the gift and passion of communication with the greatest generosity that influenced several generations of chemists who approached her or read her publications and books. Her strength of character was legendary and served a rigorous personal discipline. Above all, she always manifested an uncommon generosity in human relationships, including a great loyalty to her friends. In a word, she was a kind of genius, a *Don Quixote* of our time.

Bianka Tchoubar was of Karaite origin, a Jewish sect from Babylon that orthodox rabbis considered to be "accursed", and that still exists in Crimea. She was born in Kharkov, Ukraine, on October 22, 1910. Her family left Russia with two children in 1920, because her father was a constitutional democrat (cadet) close to Pavel Milyukov and Vladimir Nabokov. They lived for two years in Constantinople where she learnt French, then in Budapest. Finally in 1924 she arrived in Paris, where she then attended the Russian School. There, her chemistry teacher, Miss Chamier, of Russian origin in spite of her French name, was a collaborator of Marie Curie. Bianka later said that she owed her vocation to her. It is probable that Marie Curie also influenced Bianka Tchoubar, as she indicated that she had attended her courses with admiration.

She completed the "Licence ès Sciences" in 1931, then worked at the Faculté des Sciences at the Sorbonne with Paul Freundler, a close friend of Joseph-Achille Le Bel (1847–1930) with whom he shared an interest in asymmetric nitrogen. Therefore, Tchoubar worked on the reaction of ethyl iodoacetate with tertiary amines, which led her to a graduate degree (Diplôme d'Etudes Supérieures) in 1932. She then was hired by Marc Tiffeneau, Professor at the *Faculté de Médecine* in Paris, with whom she published her first article in 1934 on the reaction of Grignard reagents on α-chlorocyclohexanones. In 1937, she created her own team dedicated to organic transposition reactions, became head of the *Organic Chemistry Laboratory* and Research Trainee at the *"Centre National de la Recherche Scientifique"* (CNRS) which had just been created.

During the war, Bianka Tchoubar played a marked role in the Resistance. In 1946, encouraged by Jeanne Levy (another brilliant student of Tiffeneau) after Tiffeneau's death in 1945, she presented her doctorate thesis *Contributions to the Study of Ring Extensions: Nitrous Desamination of 1-aminomethyl, 1-cyanohexanols*, then became appointed Research Assistant in the CNRS. It was in the mid-1950s that she developed strong ties with Soviet colleagues, in particular E. A. Shilov of the Organic Institute of the Ukrainian Academy of Sciences.

Bianka Tchoubar was not appointed Director at the CNRS before 1955, although she had been a team leader for 18 years, due to her modern ideas against the current views on reaction mechanisms as well as her political opinions. In 1960, she published in French her first and famous book, *Reaction Mechanism in Organic Chemistry* that was further translated into six languages and re-edited twice. In 1961, she moved to Gif sur Yvette (near Paris) to the *Institut de Chimie des Substances*

Naturelles where her high reputation as a leader in French organic chemistry became firmly established.

In 1968, she became the Director of the *CNRS Laboratory Center No 12* that was being created with 50 researchers in Thiais (near Paris) including the teams of Micheline Charpentier, Mariann Kopp, Geneviève Le Ny, Henriette Rivière, Zoltan Welvart, and later Daniel Lefort, Jacqueline Seyden-Penne, Michel Simalty, Helena Strezlecka, Georges Bram and Paulette Viout. She actively held that position with great scientific authority for a decade until her official retirement in 1978. Her scientific activities did not stop at that stage and, in Thiais, she then became interested in organometallic chemistry and nitrogen fixation, an area that she developed with Geneviève Le Ny and Michelle Gruselle in strong collaboration with Russian friends, Professor Alexander E. Shilov, the son of E.A. Shilov, and Alla Shilova. Bianka Tchoubar published her second book in 1988 (in French) in collaboration with André Loupy, *Salt Effects in Organic and Organometallic Chemistry* that was translated into English and Russian.

During her retirement, Bianka Tchoubar's activity never diminished. Her last scientific work was a *Chemical Review* article written with her friends André Loupy and Didier Astruc on *Salt Effects Resulting from Exchange Between Ion Pairs*. Altogether, she is the author of 140 publications. She also passionately enjoyed the social, cultural and artistic life in Paris with her friends (including the author of this biography) visiting her in her old studio not far from the Eiffel Tower. Bianka Tchoubar passed away in her home on the morning of April 24, 1990, from internal hemorrhage.

Bianka Tchoubar's Fight for Modern Ideas on Organic Chemistry Mechanisms Against Conservative French Professors in the First Half of the Twentieth Century

Already at the age of 22, Bianka Tchoubar brought her interest in charged species to the attention of her Sorbonne Professor Paul Freundler, but he said to her, "if you've come to speak to me about ions, go away...When students begin giving me a lecture with interpretations based on ions, I stop them, and they get a zero".

With her PhD advisor Marc Tiffeneau, a reputed synthetic chemist, she said to him: "your migratory ability, and all these processes that interest and intrigue you can very well be interpreted in the light of the present day conceptions on the nature of the chemical bond". "I saw that he was interested, and I wrote a note on *the interpretation of the nitrous desamination involving ring extension*. Tiffeneau took my article and gave it back days later". "Miss, I cannot present to the Academy what are after all only interpretations". He spoke about it no more.

Later, after Tiffeneau's death in 1945, Tchoubar presented her thesis manuscript to Mme Pauline Ramart-Lucas, full Professor of Chemistry at the Sorbonne. "Young Lady", she said, "I have nothing against the experimental, descriptive section of your work. However, I strictly disagree with the interpretations. Get rid of them, I won't have any of it".

Tchoubar's promotion to *Directeur de Recherche* did not come easily. The CNRS Organic Committee that controlled that advancement had for several consecutive years upheld the candidates of Professor Charles Prévost who had no fondness for those mechanistic theories that Bianka defended.

Luckily, there were a few scientific colleagues who acknowledged her personal qualities, such as Professor Edmond Bauer, one of the best theoretical physical chemists at that time, and the excellent biochemist Louis Rapkine who asked Tchoubar to collaborate with him, which she did until his death in 1948. Later, Tchoubar finally became highly respected for her role as a pioneer in French organic chemistry. For instance, in 1981 she was awarded the Jecker prize of the French Academy of Sciences (at the age of 71!), although she was never looking for awards or honors.

Nancy Nouis (1947–2006) was a very close friend of Bianka Tchoubar, and this biography is dedicated to her memory.

Literature

Bianka Tchoubar published 140 research articles, the following are a selection.

Bazhenova, T.A., Lobovskaya, R.M., Shibaeva, R.P., Shilov, A.E., Shilova, A.K., Gruselle, M., Le Ny, G., and Tchoubar, B. (1983) Structure of the intermediate iron (0) complex isolated from the dinitrogen fixing system LiPh + $FeCl_3$. *J. Organomet. Chem.*, **244** (3), 265–272.

Loupy, A., Tchoubar, B. and Astruc, D. (1992) Salt effects resulting from exchange between two ion pairs and their crucial role in reactions. *Chem. Rev.*, **92** (6), 1141–1165.

Loupy, A. and Tchoubar, B. (1988) *Effets de Sels en Chimie Organique et Organométallique*, Dunod, Paris; (1992) *Salt Effects in Organic and Organometallic Chemistry*, Wiley-VCH, Weinheim.

Sources: numerous discussions with Bianka Tchoubar, and biography by Jean Jacques, in *Mechanisms and Processes in Molecular Chemistry* (Dedicated to Bianka Tchoubar), ed. D. Astruc, *New J. Chem.* 1992, **16**, 8–10, and English translation by Nancy Nouis, *New J. Chem.*, 11–13.

Tchoubar, B. (1964) Quelques aspects du rôle des solvants en chimie organique. *Bull. Soc. Chim. Fr.*, 2069.

Tchoubar, B. (1960) *Les Mécanismes Réactionnels en Chimie Organique*, Dunod, Paris, (2nd edn. 1964 and 1968); (1966) *Reaction Mechanism in Organic Chemistry*, Iliffe Books, American Elsevier Pub. Co, New York.

Tchoubar, B. (1956) Etat actuel de la théorie de la structure en chimie organique. *Nuovo Cimento*, **101**, Suppl. No 1, vol. 4, sér. X., 101.

Dorothy Crowfoot Hodgkin (1910–1994)

Renate Strohmeier

■ Dorothy Hodgkin was the third women to win the Nobel Prize in Chemistry and the last for the following 45 years.

To characterize Dorothy Crowfoot Hodgkin's scientific personality, Max F. Perutz (Nobel Prize for Chemistry 1962) wrote: "She had the courage, skill, and sheer willpower to extend the method (X-ray crystallography) to compounds that were far more complex than anything attempted before. ... Dorothy Hodgkin's uncanny knack of solving difficult structures came from a combination of manual skill, mathematical ability and profound knowledge of crystallography and chemistry. It often led her, and her alone, to recognize what the initially blurred maps emerging from X-ray analysis were trying to tell".

In the early 1940s, penicillin was isolated by Chain and Florey in Oxford and some of the best chemists tried to find its chemical constitution – with no success. Dorothy Hodgkin and her colleagues were the first scientists to use X-ray analysis, not chemistry, to determine the structural arrangement of penicillin, and succeeded in 1945. This was followed in later years by the discovery of the structure of such complex molecules as Vitamin B12 and even the protein Insulin, a thousand times larger than Vitamin B12. These findings made it possible to manufacture these vital substances and to provide undreamed-of possibilities for medical treatments of previously incurable diseases.

"I was captured for life by chemistry and by crystals" Dorothy Crowfoot Hodgkin later remembered, "when I learned how to make solutions from which to grow crystals" at about the age of ten. She went to a small private school, set up by parents of independent views. Here she attended classes in physics and chemistry, which were not in the syllabus of most elementary schools, particularly not for girls. In her own "attic laboratory" she went on with chemical experiments. At the age of 16 her sister Molly bought her two books by the scientists Bragg, father and son, who had established the use of X-rays to study the atomic structure of materials. She recalled later "I was fascinated by the way this knowledge (arrangement of the atoms) was acquired – by passing X-rays through crystals and studying the diffraction effects produced by the atoms on the X-rays. I began to see X-ray diffrac-

European Women in Chemistry. Edited by Jan Apotheker and Livia Simon Sarkadi
Copyright © 2011 WILEY-VCH Verlag GmbH & Co. KGaA, Weinheim
ISBN 978-3-527-32956-4

Dorothy Crowfoot Hodgkin

tion as a means to exploring many of the questions raised but left unanswered by school chemistry – the structure of solids and of biological materials".

She had most of her secondary education at Sir John Leman School in Beccles, Suffolk, where she and one other girl were allowed to join the boys in chemistry classes.

About the same ratio of male to female students she came upon at Oxford when she began her studies in chemistry. At that time about 10 percent of the five thousand students in Oxford were women. Particularly small was the number of women wishing to read science or mathematics. In her year, the number of women reading chemistry was amazingly high: five women in the five women's colleges. In 1933 Dorothy Crowfoot went to Cambridge to start working for her doctorate degree with J.D. Bernal. They recorded the X-ray diffraction pattern of pepsin, which was the first globular protein analyzed by this method. During that time (1934) she suffered the onset of a severe case of rheumatoid arthritis that progressively crippled her throughout the rest of her life.

To establish the penicillin structure, the first IBM analog computers were used for the X-ray calculations, and so Dorothy Hodgkin pioneered the application of electronic computers to biochemical problems.

A new problem was taken up when Lester Smith of the Glaxo pharmaceutical company asked for support in a letter in 1948: "I have recently isolated from liver, as red needle crystals, the factor that is specific for the treatment of pernicious anaemia, and we are anxious to obtain as much information as possible on its crystallographic structure. ...We wondered whether you would be sufficiently interested to undertake some X-ray measurements on the crystal..." And she was interested! To solve the structure of the crystal, vitamin B 12, took her and her coworkers eight years. The first X-ray diffraction photographs revealed that vitamin B 12 is composed of over a thousand atoms while penicillin has only 39. The vitamin contained a ring system unlike any that had been seen before. Knowledge of the structure and the constituent atoms gave clues to its function and facilitated its synthesis. As with penicillin, and some years later insulin, it had a clear clinical value. The results were published in *Nature* in 1955 and 1956.

In 1964 the Nobel Prize in Chemistry was awarded to Dorothy Crowfoot Hodgkin "For her determinations by X-ray techniques of the structures of important biological substances". For the following 45 years no other woman won a Nobel Prize in Chemistry. It was only in 2009 when Ada Yonath (born 1939 in Jerusalem) was awarded a Nobel Prize in Chemistry for her studies on the structure and function of ribosomes determined by X-ray crystallography, a physical method first extended to (bio)chemistry by her predecessor Dorothy Hodgkin.

A long persisting challenge was the analytical investigation of the structure of the hormone insulin. Eventually the advancement of computer technology contributed considerably to the ability to calculate the results. Dorothy Hodgkin first became interested in insulin in 1934 when Robert Robinson gave her a small sample to photograph. She describes the moment when she first saw "the central pattern of minute reflections" (on the photo) as "probably (the) most exciting of my life". She completed the deciphering of the three-dimensional structure of the insulin protein 35 years later, in 1969.

Since childhood, international peace and understanding had been a major concern of Dorothy Hodgkin, initially stimulated by her mother who lost her four brothers in World War I. First, her idealism manifested itself through support and encouragement for students and scientists from all over the world, regardless of whether from communist or capitalist countries. It was not before she became famous for her Nobel Prize that she began to campaign internationally for peace and disarmament. Unlike some close friends and colleagues she never joined the Communist Party but a number of associated organizations like *Science for Peace* and the *Campaign for Nuclear Disarmament*. Nevertheless, a visa application for the U.S.A was declined in 1953 and was not permitted until 1990, while the Soviets invited her and awarded her with the Mikhail Lomonosov Gold Medal of the Soviet Academy of Sciences in 1982 and the Lenin Peace Prize in 1987, amongst others. In 1976 she was elected President for some years of the "Pugwash" conferences on Science and World Affairs, conferences with "the purpose to bring together influential scholars and public figures concerned with reducing the danger of armed conflict and seeking cooperative solutions for global problems". These aims of Pugwash sound much like her personal credo. Her friend and colleague Max F. Perutz remembered her presidency: "In the face of diametrically opposed views, often angrily expressed by scientists from East and West or North and South, a few gentle thoughtful words in her soft voice cooled tempers and forestalled crises".

Literature

Cochran, W. (1996) Dorothy Mary Crowfoot Hodgkin, OM, FRS. *The Royal Society of Edinburgh Year Book*, Session 1994–1995.

Cohen, L.J. (1996) *Dr. Dorothy Crowfoot Hodgkin: Chemist, Crystallographer, Humanitarian (1910–1994).* http://nobelprizes.com/nobel/chemistry/dch.htlm

Dodson, G, Glusker, J.P. and Sayre D. (Eds.) (1981) *Structural Studies on Molecules of Biological Interest: A Volume in Honour of Professor Dorothy Hodgkin*, The Clarendon Press, Oxford.

Ferry, G. (1998) *Dorothy Hodgkin, A Life,* Granta Books, London

Fölsing, U. (1994) Dorothy Hodgkin-Crowfoot, Chemie-Nobelpreis 1964 in *Nobel-Frauen. Naturwissenschaftlerinnen im Porträt*, Beck, München.

Glusker, J.P. and Adams, M.J. (1995) Dorothy Crowfoot Hodgkin (1910–1994). *Physics Today*, May 1995

Perutz, M.F. (1995) Dorothy Crowfoot Hodgkin. Crystallographers Online

Ulla Hamberg (1918–1985)

Carl G. Gahmberg and Pekka Pyykkö

Ulla Margareta Hamberg (20 October, 1918–22 March, 1985) was one of the most notable Finnish biochemists of her lifetime, an intrepid pioneer of international cooperation, and one of the early students of the peptide bradykinin.

What made Hamberg remarkable was the length and standard of her foreign and domestic collaborations. Eight of her early papers from 1948 to 1953 were written while she was an assistant to Professor U.S. von Euler (1905–1983) (later Nobel-Prize winner in Medicine 1970) at the Karolinska Institute at Stockholm. She then moved as Research Associate to São Paulo, Brazil, working at the Instituto Biologico and the Faculty of Medicine of the University of São Paulo from 1954 to 1958. Her most important coauthor was Professor M. Rocha e Silva. A previous stint at the University of Wisconsin from 1956 to 1957 with H.F. Deutsch was followed by another stay in the United States at Cleveland with I.H. Page from 1959 to 1961. She

Ulla Hamberg (web exhibition "Women of Learning"; http://www.helsinki.fi/akka-info/tiedenaiset/ english/hamberg.html).

European Women in Chemistry. Edited by Jan Apotheker and Livia Simon Sarkadi
Copyright © 2011 WILEY-VCH Verlag GmbH & Co. KGaA, Weinheim
ISBN 978-3-527-32956-4

was connected with the University of Helsinki from 1959 and published papers with A.I. Virtanen (1895–1973) (Nobel-Prize winner in Chemistry 1945) and later with A. Vartiainen and J. Erkama. From 1965 she was an independent scientist, holding mostly positions as Research Scientist of the National Research Council for Sciences from 1966 until 1976. From June 1976 until her death she held a personal professorship of biochemistry at the University of Helsinki. She had previously obtained a tenured position as Associate Professor of Biochemistry at the University of Turku in 1967 but resigned less than two years later.

In 1947, she published her first article with Virtanen on the removal of amide groups from plant proteins. Virtanen had received the Nobel Prize in chemistry in 1945, and his main interest was in plant biochemistry and agricultural applications. But Ulla had other plans. She left the Virtanen laboratory and went to the Karolinska Institute in Stockholm to learn modern pharmacology and biochemistry. She was fortunate to join the group of U.S. von Euler, a world-famous biochemist working on neural transmitters. von Euler had discovered noradrenaline and shown its importance in signaling in the nervous system. Another group member was Sune Bergström, who later discovered prostaglandins and, like von Euler, became a Nobel laureate.

Ulla Hamberg's main contributions from this time were the development of useful assays for noradrenaline and its separation from adrenaline (epinephrine). She also showed that noradrenaline is enriched in the adrenal medulla. At that time, Scandinavian scientists used to publish in Scandinavian journals, and her most cited work on the analysis of noradrenalin appeared in *Acta Physiologica Scandinavica*. However, she also published in *Nature, Science,* and *the Biochemical Journal*. Due to this important work, she became well known. She wanted to continue with protein chemistry and joined the laboratory of the Brazilian scientist M. Rocha e Silva in the early 1950s. He was famous because of his discovery of bradykinin, a nine-amino acid peptide found in human plasma.

Bradykinin has strong pharmacological effects, lowering the blood pressure due to dilatation of the blood vessels. Using a snake venom, she showed that bradykinin is released from a higher-molecular-weight protein in plasma. This finding turned out to be important. She also found that the proteolytic enzyme trypsin had a similar effect. Bradykinin originates from kininogens, and it became important for her

Bradykinin.

to continue studies on functionally important plasma proteins. After her stay in Brazil, therefore, she began to study plasminogen and its activation. Plasminogen is the immediate precursor of plasmin, and it is formed by proteolytic cleavage. She isolated urokinase and used streptokinase-activated plasma to induce plasminogen activation (cleavage). Plasmin has an important role in fibrinolysis and the mechanism and the regulation of its activity is of pivotal importance.

In several articles, she characterized the activation of plasmin and the formation of large protein complexes in plasma.

She later returned to bradykinin research and published several papers on kinins and kininogen and their regulation.

Her early work with von Euler remains her most important contribution. She had a background in pharmacology, and, throughout her career, she remained faithful to pharmacologically active substances and, notably, plasma proteins and peptides derived from them. She was highly respected by her contemporaries, but during her later research period she did not keep up with the developments in protein chemistry. Her work remained rather descriptive, and the molecular mechanisms of peptide formation turned out to be difficult for her to solve. She established, however, a good group of graduate students working on plasma proteins, and with the limited financial resources she had, her scientific contributions are remarkable.

Few Scandinavian women scientists of her time became as well known, and she certainly acted as a pioneer in the development of biochemistry in Finland. Importantly, she showed that a woman scientist can get far in research, and this fact has certainly inspired many female students in our country to take up a scientific career.

In her will, to the Finnish Society of Sciences and Letters, of which she had been an active member, she specified that grants should be given to cancer research.

Literature

Bergström, S., von Euler, U.S. and Hamberg, U. (1949) Isolation of nor-adrenaline from the adrenal gland. *Acta Chemica Scandinavica*, **3** (3), 305–305.

Hamberg, U. and Silva, M.R.E. (1957) Release of bradykinin as related to the esterase activity of trypsin and of the venom of bothrops-jararaca. *Experientia*, **13** (12), 489–490.

Hamberg, U. and Silva, M.R.E. (1957) On the release of bradykinin by trypsin and snake venoms. *Archives Internationales de Pharmacodynamie et de Therapie*, **110** (2–3), 222–238.

Virtanen A.I. and Hamberg U. (1947) On the splitting of the amide group from proteins – the amides of zein. *Acta Chemica Scandinavica*, **1**(9), 847–853.

von Euler, U.S. and Hamberg, U. (1949) Colorimetric estimation of noradrenalin in the presence of adrenalin. *Science*, **110**, 561–561.

von Euler, U.S. and Hamberg, U. (1949) Colorimetric determination of noradrenaline and adrenaline. *Acta Physiologica Scandinavica*, 19(1), 74–84.

von Euler, U.S. and Hamberg, U. (1949) L-noradrenaline in the suprarenal medulla. *Nature* **163**, 642–643.

Rosalind Franklin (1920–1958)

Marianne Offereins

Rosalind Franklin was one of the discoverers of the structure of the DNA molecule. She provided the experimental data which formed the basis of the research for which Watson, Crick and Wilkins were granted the Nobel Prize in 1962.

On July 25, 1920 Rosalind Elsie Franklin was born in London, the second of five children of the wealthy Jewish banker, Ellis Franklin and his wife Muriel Waley.

Because she did not like specific girl games, she was considered an outsider, so she remembered her childhood as an ongoing struggle for recognition.

When Rosalind was eight years old, she was sickly and often had respiratory problems so the family physician advised to send her to a boarding school near the

Rosalind Franklin (http://sciencecomm.wikispaces.com/ file/view/3441067.jpg/96607078/3441067.jpg).

European Women in Chemistry. Edited by Jan Apotheker and Livia Simon Sarkadi
Copyright © 2011 WILEY-VCH Verlag GmbH & Co. KGaA, Weinheim
ISBN 978-3-527-32956-4

sea. The lesson she learned there was to better ignore pain and disease. In London she attended St. Paul's Girls' School, a high school for upper class girls. While at St. Paul's, she spent part of a semester in Paris. She came back as a, very fashionable, convinced Francophile; from then on she made her own clothes, according to French fashion.

Especially, the science education at her school was excellent, and at the age of 15 her decision was made: she wanted to study physical chemistry at Cambridge. In 1938 Rosalind Franklin began her studies at Newnham College, Cambridge. In a letter home she described it as: "Rather like a boarding school". She worked hard and was dedicated, and, in 1941, she graduated. After that she worked for a year in a research function with the future Nobel laureate Ronald Norrish.

It was the Second World War and Rosalind wanted to serve her country, therefore, in 1942, she went to work as an assistant research officer at the 'British Coal Utilisation Research Association' (CURA), an organization doing research on ways to improve the use of coal as a fuel. There "she did fundamental research in the area of coal and graphite microstructures." Until 1946 she was employed by CURA and in this period she obtained her doctorate for her research on gas chromatography (in 1945). One of her professors said about her research, "she [...] brought order into a field which had previously been in chaos". During her research at CURA, she published five papers which are still regularly cited.

Her reputation was made, but Rosalind was ready for a new challenge. In 1947 she traveled to France, where she had found a job at the *Laboratoire Central des services Chimique de l'Etat*. Here she specialized in X-ray diffraction. During this time Rosalind developed great skill in performing X-ray analytical research of amorphous substances, which are clearly not crystals, so X-ray crystallography could be suitable for the research on the structure of DNA. She became the expert in analyzing X-ray photos of substances on the border of crystalline and non-crystalline, which means research on carbon and 'biomolecules'.

According to friends, these were the happiest years of her life.

In 1950, she returned to London at the request of Sir John Randall, the director of King's College (not to be confused with King's College, Cambridge), who assigned her as a research associate to set up DNA research in the laboratory. He transferred all research to her, "Lock, stock and barrel".

In the meantime, in Cambridge, James Watson and Francis Crick had started their research on the same subject, as did Linus Pauling in America. At first, the investigations in London and Cambridge went more or less at the same level. The first article Rosalind wrote about DNA appeared in the same issue of *Nature* as the article by Watson and Crick.

Soon Rosalind could demonstrate that the DNA molecules appear in two forms: A and B, depending on the amount of water they absorb: a relative humidity of 75% leads to the A-shape, a humidity of 95% leads to the B-form. The B is 25% longer and the coils are stretched. She could make the molecules change from one form to another by varying the humidity. Because the molecules could so easily absorb moisture from the surrounding air and exude it, she deduced the place of the sugar-phosphate components which were known to occur in DNA. She also conclud-

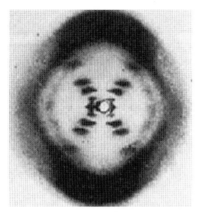

X-ray diffraction pattern from DNA
(http://edu.glogster.com/media/1/9/12/15/9121555.jpg).

ed that the phosphate backbone was on the outside of each chain, the organic bases were on the inside to form the steps of the 'stair'. At that time it was not clear to her whether each molecule consisted of two, three or four chains.

In May 1952 Rosalind made an X-ray of the B-form of DNA. This picture clearly shows an X-shaped pattern, suggesting that the B-form is a helix (spiral). At that time she did not pay much attention to the photograph, and the X-ray disappeared temporarily into a drawer because she first wanted to calculate the complete A-form.

Typical behavior for her, because that was the logical order. From the calculations of A, she subsequently obtained a lot of data for B.

In the meantime, Watson and Wilkins became good friends, and when the two men discussed the issue of DNA, Wilkins showed Watson the X-ray on January 30, 1953, without Franklin's permission and without even telling her! The photo gave proof of the theory Watson and Crick in Cambridge had begun to set up. A few weeks later, Max Perutz, head of the laboratory in Cambridge, received a government report which contained all the essential information from Franklin's research. Without any permission from Randall or Franklin, he passed it on to Crick. Now all vital data were available to Cambridge. Later, Watson and Crick would get all the credits, the name Franklin was not even mentioned.

Rosalind, however, never knew how her work led to the triumph of Watson and Crick. At the time of their publication, she actually no longer worked on DNA, because in March, 1953 she moved to Birkbeck College, where she focused her research on the tobacco mosaic virus (TMV) and the virus that causes polio. She showed that the carrier of the genetic characteristics of the virus, the RNA, like DNA was a helix, and a model of its structure TMV was shown in 1957 at the Brussels World Exhibit.

Between 1953 and 1958 she had 17 publications on viruses and she laid the foundation for structural virology.

During her first three years at Birkbeck everything she did was, according to her biographer Anne Sayre, excellent. The contact with the group from Cambridge was friendly and they exchanged data concerning the field of viruses. In 1956 the tide turned. Her grant was withdrawn, because the grant giver did not want to spend money on a project led by a woman; from that moment she received the money for her research from the U.S. Public Health Service. In the summer of that year, during a trip to America, for the first time she suffered from such severe abdominal pain that she had to find a physician. He advised her, to visit a specialist in England as soon as possible. After consulting him, it became clear that she had ovarian cancer. All treatments failed and she died on April 16, 1958. She was only 37 years old.

In 1962 Watson, Crick and Wilkins received the Nobel Prize for their research into the structure of DNA. In their Nobel lectures they have 98 references, not one to Franklin.

J.D. Bernal, head of the Birkbeck laboratory, said about her: "As a scientist Miss Franklin was distinguished by extreme clarity and perfection in everything she undertook. Her photographs are among the most beautiful X-ray photographs ... ever taken".

Acknowledgement

Thanks to Dr. Leo Molenaar, Dr. Sheila Tobias and Dr. Ep Hartman.

Literature

Kass-Simon, G. and Farnes, P. (eds.) (1993) *Women of Science, Righting the Record*, Indiana University Press, Bloomington.

McGrayne, S. (1993) *Nobel Prize women in science*. Birch Lane Press, New York.

Rozendaal, S. (1998) *De mens, een dier. Denkers aan het front van de wetenschap*. De Bezige Bij, Amsterdam.

Sayre, A. (1978) *Rosalind Franklin and DNA*. W.W. Norton, New York / London.

Simmons, J. (1997) *De top-100 van Wetenschappers. De 100 Meest Invloedrijke Weten-schappers uit Heden en Verleden op een Rij Gezet* (The Scientific 100), Het Spectrum, Utrecht.

Sluyser, M. (1998) *Waarom kreeg Rosy geen Nobelprijs?* Vrij Nederland 15 August, 1998.

Watson, J.D. (1968) The Double Helix. *A Personal Account of the Discovery of the Structure of DNA*. Athenaeum, New York.

Yount, L. (1996) *Twentieth Centurywomen-scientists*. Facts on File Inc., New York.

Jacqueline Ficini (1923–1988)

Jean-Pierre Genet

■ Jacqueline Ficini's research interest focused on the fundamental organic reactions and their applications in synthetic organic chemistry, a domain she regarded as endlessly fascinating in itself. She was best known for her initiation and development of ynamine chemistry.

Born in 1923, October 20, at Saint-Maixent-l'École (Deux-Sèvres, France), daughter of Jane Pontet and Colonel Raoul Ficini. Jacqueline Ficini, single, died in 1988 in Paris when returning from Japan where she had visited Professor Yoshida at his invitation.

Jacqueline Ficini completed her secondary education at the Collège Sainte-Marie de Chavagne at Angoulême, and got her Master's Degree at the Universities of

Jacqueline Ficini (1923–1988)
© Jean-Pierre Genet, private collection.

European Women in Chemistry. Edited by Jan Apotheker and Livia Simon Sarkadi
Copyright © 2011 WILEY-VCH Verlag GmbH & Co. KGaA, Weinheim
ISBN 978-3-527-32956-4

Paris and Angers. After her PhD (1952) in industry with Dr. R. Rothstein as supervisor, she was appointed research assistant at CNRS from 1952 to 1956. In 1957 she became instructor in chemistry at the Faculty of Science in Paris where she worked with Professor H. Normant.

In 1960 she spent a postdoctoral year at the Columbia University with G. Stork as a Research Associate. In 1962, when she returned, she moved to the Faculty of Science in Reims as Maître Assistant, where she was also promoted Maître de Conférence. She was promoted Full Professor at the University of Paris VI in 1965. She assumed several responsibilities at the University of Paris VI, where she headed a doctoral school in organic chemistry. Under her direction, from 1964 to 1988, 18 PhD (doctorat d'état) and 26 university thesis or engineer thesis were defended; 6 research associates (CNRS) and 8 assistant professors were with her in her laboratory.

She was elected President of the Organic Chemistry Division at the French Chemical Society where she was rewarded with the Le Bel Prize for her contribution in the field of ynamine chemistry in 1972. Ficini's achievements have been recognized by the French Academy of Sciences with the Jecker Prize and Berthelot Medal in 1979. She also received honors such as the Palmes Académiques (French decoration for services to education, 1974), and the Legion of Honour (1986). As member of the American Chemical Society, she was in constant contact with the international community; she was also several times Invited Professor at Columbia University New York (1975, 1978, 1985).

She gave about 130 Invited Lectures at National and International Symposia and published over 120 articles, including reviews and patents. She was also very interested in teaching. In 1968, she published *Structure de la Matière et Cinétique Chimique* with N. Lumbroso-Bader and J.C. Depezay, a successful text book dedicated to students beginning chemistry at the university, republished in 1976, 1981 and 1986. This book is still in libraries and bookstores.

Jacqueline Ficini had a sort of panache that generated enthusiasm among her group. She was full of ideas, her curiosity to explore synthetic organic chemistry was captivating to her listeners and it was a truly marvelous opportunity to have been under her direction. There was a time in the 1970s and 1980s when the group was reasonable in size with a good balance between PhD students and permanent researchers. This was the time of C. Barbara, J. d'Angelo, J. Besseyre, M. Claes, J.C. Depezay, D. Desmaele, A. Dureault, J.P. Genet, A. Guingant, P. Kahn, A. Krief, J. Pouliquen, G. Revial and A.M. Touzin. Her students found high-level employment in industry or research centers in: USA, France (CNRS), Belgium, UK, Switzerland. Others became successful professors in Belgium and in France.

Her interests in chemistry were wide. At first the main area of her work was the investigation of organometallic compounds, the synthesis and reactivity of Grignard's reagents with Professor Henri Normant in 1956. In 1961, with Professor Gilbert Stork, she developed the first catalyzed intramolecular cyclization of an unsaturated diazocarbonyl compound that formed a cyclopropane. These catalysts are still used today. Later, at the University of Paris VI, she developed a lithium-halogen exchange reaction between functionalized vinyl halides and alkyl lithium as base. This procedure offered useful functionalized reagents. Professor Ficini also developed general methods for the synthesis of small rings such as cyclobutenones, and cyclopentenones. For instance, these original methods have been used in the preparation of cinerolone and jasmolone, key components of pyrethrins. Ficini was also attracted by transition-metal catalyzed reactions. An original chemo- and regioselective intermolecular alkylation of bifunctionalized allylic acetates using Pd(0) catalysts was discovered. A new cyclopropanation route to vinyl cyclopropanes was developed via π-allyl palladium chemistry and used in the synthesis of chrysanthemic acid, a key component of pyrethrinoids, natural and biodegradable insecticides. Simultaneously, iron-catalyzed cycloadditions of ynamines where developed by her group. A very convenient *in situ* generated iron(0) complex: $FeCl_3/i\text{-PrMgCl}$, in the presence of butadiene and ynamines, afforded the cyclohexadienamines.

chrysanthemic acid

Jacqueline Ficini was best known for her initiation and development of ynamines chemistry. After the establishment of their practical synthesis, she demonstrated that ynamines also react rapidly with alcohols leading to *O*-, *N*-acetals. In the case of allylic and propargylic alcohols the initial adducts undergo Claisen rearrangement with formation of amides, known as the *Ficini–Claisen rearrangement*.

This rearrangement can be considered as complementary to the Eschenmoser rearrangement. Ynamines are basic enough, in contrast to acetylenic ethers, to enolize carbon acids. For example, the reaction of ynamines with 5-membered enol lactones leads to enaminolactones. These enaminolactones, after hydrolysis, provide a new route to 1,4 diketones. This method has been applied to the synthesis of jasmone, a useful perfume.

Acylation of ynamines by bicyclic enol-lactones using $MgBr_2$ as a mild Lewis acid catalyst provides an efficient method of spiroannelation leading to spiro (4.5)

jasmone

decenes and spiro (5.5) undecenes. This was used in the synthesis of (D, L) acoradiene, a sesquiterpene present in the oil of *Vetiveria zizanoides.*

She also demonstrated that the ynamines have a special propensity towards cycloadditions. Ynamines react with great facility with a heterocumulene, such as car-

acordiene

bon dioxide, giving access to allenic diamides. In addition to this remarkable reaction, different types of cycloadditions of ynamines have been discovered using electrophilic substrates: acetylenes, dienes, imines, isocyanates, ketenes, unsaturated ketones. Ficini's contribution is outstanding, for instance, in the cycloaddition of ynamines with enones having a cisoid conformation the (2+4) process occurs, leading to heterosubstituted pyranes. On the other hand cycloaddition of the (2+2)-type takes place stereoselectively with transoid enones such as cyclohexenones and cyclopentenones, giving bicyclic cycloadducts ynamine, which after hydrolysis gives keto acids. This two-step sequence allows perfect control of the relative configuration of two contiguous carbon centers via the formation of the kinetically preferred stereoisomer. This represents a unique stereoselective route to diastereoisomeric five- and six-membered 1,5 keto-acids. Ficini's ynamine-cycloaddition cleverly controls the relative stereochemistry at C_3, C_{15}, and C_{20} in dihydroantirhine synthesis. The stereocontrolled synthesis of Des-AB-cholestane and cholestene, key units for building vitamin D3 hydroxylated metabolites, has been achieved using this powerful method. The first stereocontrolled synthesis of juvabione also uses this elegant chemistry. Juvabione exhibits juvenile hormone activity in insects. It should be noted that this synthesis is described in a textbook, *Advanced Organic Chemistry,* by Carey and Sundberg.

For 20 years, the synthetic significance of ynamines in organic and organometallic chemistry was firmly established by Ficini's contribution as well as that of Pro-

juvabione

fessor Heinz G. Viehe at the University of Louvain (Belgium). These elegant pioneer works were informatively and carefully reviewed by Ficini in *Tetrahedron* in 1976. Applications of the original and versatile chemistry allowed Ficini to achieve truly efficient syntheses of biologically active molecules.

The chemistry of ynamines has re-emerged in the last five years in the form of ynamides, revitalizing interest in these functionally rich building blocks. The first stereoselective *Ficini–Claisen* rearrangement using chiral ynamides has been reported very recently, 36 years after Ficini's discovery. This clearly shows the impact of her work on modern synthetic organic chemistry.

Literature

Carey, F. A. and Sundberg, R. J. (2007) *Advanced Organic Chemistry*, 5th edn., vol. B, Plenum Press, New York.

Depezay, J. C. and Ficini, J. (1968) Formation, stabilité et utilisation en synthèse d'organovinyl-lithiens d'éthers β bromés et chlorés. *Tetrahedron Lett.*, **9** (8), 937–942.

Ficini, J. (1976) Ynamine: A versatile tool in organic synthesis. *Tetrahedron*, **32** (13), 1449–1486.

Ficini, J. and Barbara, C. (1964) A general synthesis of ynamines. *Bull. Soc. Chim. Fr.*, 871.

Ficini, J. and Krief, A. (1970) Stereochemical control in the hydrolysis of an ynamine-cyclopentenone adduct. *Tetrahedron Lett.*, **11** (17), 1397–1400.

Ficini, J. and Pouliquen, J. (1971) Cycloaddition of ynamines with carbon dioxide. Route to diamides of allenes-1,3 dicarboxylic acids. *J. Am. Chem. Soc.*, **93** (13), 3295–3297.

Ficini, J., d'Angelo, J. and Noiré, J. (1974) Stereospecific synthesis of D,L juvabione. *J. Am. Chem. Soc.*, **96** (4), 1212–1214;

Ficini, J., Desmaele, D., Touzin, A.M. and Guingant, A. (1983) Synthèse totale et stéréosélective du *p*-tolylsulfonylméthyl-8

Des-AB-cholestene. *Tetrahedron Lett.*, **24** (30), 3083–3086.

Ficini, J., Guingant, A. and d'Angelo, J. (1979) A Stereoselective Synthesis of (+/-) dihydroantirhine. *J. Am. Chem. Soc.*, **101** (5), 1318–1319.

Ficini, J., Lumbroso-Bader N. and Depezay J.C. (1968–1969) *Éléments de Chimie Physique*. I. *Structure de la Matière et Cinétique Chimique*; II. *Thermodynamique. Équilibres Chimiques*, Hermann, Paris.

Ficini, J., Piau, F. and Genet, J. P. (1980) A novel synthesis of (+/-)-trans-chrysanthemic acid. *Tetrahedron Lett.*, **21** (33), 3183–3186.

Genet, J.P. and Ficini J. (1979) Cycloaddition des ynamines avec le butadiène catalysé par le fer (0): Synthèse de cyclohexadiènamines-1,4 et de cyclohexènones β,γ et α,β insaturées. *Tetrahedron Lett.*, **20** (17), 1499–1502.

Selected papers of Ficini among the most recently cited:

Stork, G. and Ficini, J. (1961) Intramolecular cyclization of unsaturated diazoketones. *J. Am. Chem. Soc.*, **83** (22), 4678.

Andrée Marquet (1934–)

Danielle Fauque and Andrée Marquet

Andrée Marquet spent her career both in research and teaching in organic and bioorganic chemistry. She focused her research on organic reaction mechanisms, before specializing in the understanding of biochemical processes, with a special interest in the biosynthesis of biotin (vitamin H). She was one of the founders of the bioorganic community in France.

She took responsibilities in many committees at CNRS (National Center for Scientific Research), and also at the *"Ministère de l'Education Nationale, de la Recherche et de la Technologie"* (1998). As an emeritus professor since 2000, she now devotes much time to "Chemistry and Society", as president of this commission at the *Fondation de la Maison de la Chimie*.

On March 3, 1934, Andrée Marie Marguerite Marquet was born in Tilchatel (Côte d'Or) from a family of farmers. Very soon attracted by organic chemistry, from the

Andrée Marquet (1934–), private collection © A. Marquet

European Women in Chemistry. Edited by Jan Apotheker and Livia Simon Sarkadi
Copyright © 2011 WILEY-VCH Verlag GmbH & Co. KGaA, Weinheim
ISBN 978-3-527-32956-4

secondary school, she entered the *École Nationale Supérieure de Chimie de Paris* (ENSCP) from which she received an engineering degree (1956). She was admitted, immediately after, as a member of CNRS (National Centre of Scientific Research), in Professor Alain Horeau's laboratory at *Collège de France*. She defended a thesis (1961), under Jean Jacques's (research director at CNRS) supervision. After a post-doctoral period, at the ETH in Zürich, in Professor Duilio Arigoni's laboratory, she came back to the *Collège de France* and then moved to a CNRS laboratory (1974). Her career continued at CNRS as *Maître de Recherches* (1968) then *Directeur de Recherches* (1976). She was then appointed, as full Professor, at the University P. & M. Curie, Paris (1978), where she founded the *Laboratoire de Chimie Organique Biologique*. Indeed, after studying organic reaction mechanisms for several years, she had turned to enzyme mechanisms and became a specialist in mechanistic enzymology.

She participated actively in the French scientific politics, in several councils at CNRS and the University, she was also the Director of the Chemistry Department at the "*Ministère de l'Education Nationale, de la Recherche et de la Technologie*" (1998). She has also been President of the *Division de chimie organique* of the *Société Chimique de France* (1984–1986).

Amongst honors she received the *Chevalier de la Légion d'Honneur* (1996), *Officier de l'Ordre National du Mérite* (2000), *Officier de l'Ordre des Palmes académiques* (2006), the Silver Medal of the CNRS (1988), two prizes from the French Academy of Science (both in 1986), and two prizes from the *Société Chimique de France* (1971 and 1994). She was elected corresponding member of the French *Académie des Sciences* of the *Institut de France* in 1993.

She became an emeritus Professor in 2000, and continues her activities as President of the Commission *Chimie et Société* that she helped to found within the *Fondation de la Maison de la Chimie*, in 2001, with Pierre Potier, President of the Foundation. She is also a member, since 2007, of the *Comité d'éthique du CNRS*, a pluridisciplinary structure whose mission is to publish reports on the ethics of research and on the social responsibility of scientists. A report on "the role of the scientific community in the debate on chemicals", in relation with the REACH regulation, was published in September 2009.

Andrée Marquet started her career as a researcher, in 1956, at *Collège de France*, under Jean Jacques supervision, in the *Laboratoire de Chimie Organique des Hormones*, directed by Alain Horeau. She was strongly influenced by this environment, more or less marginal with respect to the traditional University chemistry, and more open to the new concepts emerging on the international stage, but more slowly in France. She can tell us how she lived this fascinating period when organic chemistry passed from a descriptive science to a more rational one, when reaction mechanisms were becoming popular and operational, when stereochemistry became an intrinsic part of organic chemistry, with the birth and development of conformational analysis and its use in understanding the reactivity of molecules. She can tell us about the influence on the young generation of chemists of people like Bianca Tchoubar and Marc Julia with their cult books on reaction mechanisms, Guy Ourisson, who founded the GECO (*Groupe d'Etudes de Chimie Organique*), an annual meeting which played an important role in the modernisation of French organic chemistry. She remembers the famous lectures of Alain Horeau at *Collège de France*, attended on Saturday morning by the "new chemists", where the very beginning of asymetric synthesis was taught.

For Andrée Marquet, chemistry is a fantastic tool to understand nature and its laws. Of course, it is also a powerful means to transform nature and create new objects. She was, however, more interested in the first aspect, and spent her career answering the question: How does it work?

Her interest in the study of reaction mechanisms started during her PhD work, when she tried to interpret the selectivity of halogenation of ketones using phenyl-trimethylammonium-tribromide (PTT) in molecules containing very nucleophilic aromatic rings, as well as the influence of the reaction conditions on the orientation of enolization of dissymetric ketones.

Then she was busy with the study of carbanions of sulfoxides, in particular of their hybridation state and of the stereochemical aspects of the reactions in which they were involved, which were at that time the subject of controversial debates. After a spectroscopic study of the intermediate organometallic species, she proposed a unifying theory, and applied these results to a new total synthesis of biotin, a vitamin in which she became interested.

In the late 1970s, reaction mechanisms were becoming less popular, and, indeed, they were dealing with more and more "narrow" problems. On the other hand, biochemistry was offering a fascinating field of investigation for organic chemists, a field deserted at that time by many classical enzymologists moving to cell biology.

Thus, she turned to this new field that she had already discovered during her postdoctoral work, where she worked on the biosynthesis of terpenes. As she did in the field of organic chemistry, where she studied reactions important for organic synthesis, she selected in biochemistry, enzymatic reactions, which were not only an intellectual enigma from the organic chemistry point of view, but also because elucidation of their mechanisms was of interest in biotechnology or pharmacology. She worked in several areas, the mechanism of action of vitamin K, the design of inhibitors of aldosterone biosynthesis, and biotin biosynthesis.

One can point out two main characteristics in the career of Andrée Marquet, a solid autonomous scientific profile, and a constant active opening to the collective life of the scientific community.

She developed her own original research, not necessarily following the 'in vogue' areas, not especially fascinated by official recognition, but she was quietly happy to receive this recognition which followed. In a laboratory oriented to the study of stereochemical problems, and although well integrated and impregnated with this culture, she always followed her own interests, which moved from organic reaction mechanisms to biochemical mechanisms. She never specialized in a narrow field, but always made use of all available techniques, through collaborations, to solve important biological problems. She was a champion of interdisciplinarity at the time this concept was just emerging, promoting the chemistry/biology interface. She was a precursor in initiating, with a few others, national and international scientific meetings or groups, some of them still alive, centered on Bioorganic Chemistry.

She was very keen to share her knowledge with others, especially with the students. Many of them remember her lectures and quote that they were determinant for their orientation. The members of her research group have to be closely associated with the achievements reported here, which are the results of a team effort. It is unfortunately not possible to cite all her coworkers in this biography, only some of them appear in the selected references. She facilitated the autonomous initiation of new projects by the senior scientists in her laboratory, and is very happy for their present success and recognition.

At the same time, she felt highly concerned by her responsibility as a scientist, not only toward the chemists' community, but also toward society. Convinced by the importance of democratic debates of a scientific background for the citizens, she created *"Chimie et Société"* whose mission is to popularize chemistry, but also to try to understand why chemistry is so criticized by public opinion. The philosophy of this commission is that it is not sufficient to "educate" the public, but that it is necessary for a constructive dialog to take account of its feeling, its expectations and also its experience.

Literature

Bory, S., Luche M.J., Moreau, B., Lavielle, S. and Marquet, A. (1975) Une nouvelle synthèse totale de la biotine. *Tetrahedron Lett.*, **16** (10), 827–828.

Chassaing, G. and Marquet, A. (1978) A ^{13}C NMR study of the structure of sulfur-stabilized carbanions. *Tetrahedron*, **34** (9), 1399–1404.

Dubois, J., Gaudry, M., Bory, S., Azerad, R. and Marquet, A. (1983) Vitamin K-dependent carboxylation. Study of the hydrogen abstraction stereochemistry with gamma-fluoroglutamic acid-containing peptides. *J. Biol. Chem.*, **258**, 7897–7899.

Eastes, R.E. and Kleinpeter, Éd. (eds) (2008) Andrée Marquet, in *Comment Je Suis Devenu Chimiste*, Le Cavalier Bleu, Paris, pp.155–168.

Gaudry, M. and Marquet, A. (1970) Énolisation des cétones dissymétriques. Accès facile aux bromométhylcétones par bromation en présence de méthanol. *Tetrahedron*, **26** (23), 5611–5615.

Institut de France (2008) Andrée Marquet, in *Répertoire Biographique, Membres et Correspondants de l'Académie des Sciences*, Institut de France, Paris, pp. 555–556.

Marquet, A. (2010) Biosynthesis of Biotin, in *Comprehensive Natural Products II Chemistry and Biology*, (Mander, L., Lui, H.-W, eds), Elsevier, Oxford, vol. 7, pp. 161–180.

Marquet, A. (2001) Enzymology of carbon-sulfur bonds formation. *Current Opinion in Chemical Biology*, **5**, 541–549.

Marquet, A., Tse Sum Bui, B., Smith, A.G. and Warren, M. J. (2007) Iron-sulfur proteins as initiators of radical chemistry. *Nat. Prod. Rep.*, **24**, 1027–1040.

Viger, A., Coustal, S., Perard, S., Piffeteau, A. and Marquet, A. (1989) 18-substituted progesterone derivatives as inhibitors of aldosterone biosynthesis. *J. Steroid Biochem.*, **33**, 119–124.

Website of "Chimie et Société" http://www.maisondelachimie.asso.fr/chimiesociete/

Website of the CNRS ethics committee http://www.cnrs.fr/fr/organisme/ethique/comets/index.htm

Anna Laura Segre (1938–2008)

Marco Ciardi and Miriam Focaccia

■ An internationally renowned scientist recognized mainly for her contribution to understanding the polymerization mechanism of olefins through nuclear magnetic resonance. Among the earliest Italian researchers to study NMR, she became an expert in this investigative technique and, after finding innovative applications to polymers, she moved to applications in the field of foodstuffs and then cultural heritage.

Anna Laura Segre was born in Novara in 1938. She graduated in Physics at the University of Milan, academic year 1960/61, and took a PhD degree in Molecular Spectroscopy. She became Assistant Researcher in 1964, Researcher in 1967 and then Chief Researcher, always at the *Istituto di Chimica delle Macromolecole*.

She started her research in the field of Macromolecules with Giulio Natta's research group in Milan. Her input helped to clarify the polyolefin microstructure and, therefore, to better understand the polymerization mechanism of the olefins. She loved this field of research and, for this reason, in the last 15 years of her life, she worked jointly with Busico of the University of Naples.

In 1968 she was awarded a grant from the *Accademia dei Lincei*, which allowed her a stay of two years at the Carnegie-Mellon University in Pittsburgh. There she

Anna Laura Segre (Consilio Nationale delle Ricerche).

European Women in Chemistry. Edited by Jan Apotheker and Livia Simon Sarkadi
Copyright © 2011 WILEY-VCH Verlag GmbH & Co. KGaA, Weinheim
ISBN 978-3-527-32956-4

worked under the supervision of Professor Salvatore Castellano on the structure of substances oriented in mesophases.

In 1978 she moved to the Institute of Chemical Structure of CNR in the Area of Research of CNR in Montelibretti (Rome). In 1989, she was the winner, at a National level, of a position as *Dirigente di Ricerca* (Research Director) of the CNR in Macromolecular Chemistry and became the Scientific Responsible of the NMR laboratory of the Institute of Chemical Methodologies of CNR in Montelibretti (Roma).

In 1999 she moved to the Institute of Nuclear Chemistry, which became the *Istituto di Metodologie Chimiche* (Institute of Chemical Practices) in 2002. From 2001 to 2006 she was professor of Radiation Chemistry at the Pharmacology Faculty of the *La Sapienza* university in Rome.

Anna Laura Segre was a person endowed with great intellectual and scientific curiosity; thanks to this she succeeded in applying NMR to many different fields such as aromatic polymers acting as oxygen scavengers, liquid crystals, gels, the structure of tritium-labelled gas oriented in the nematic phase, the structural definition of natural and synthetic polymers in solution as well as in the solid state, and polymeric networks, structural characterization of clays, and, more recently, in food chemistry and in the chemistry of cultural heritage. The application of new NMR techniques in the field of cultural heritage was one of her many interests, and she promoted and took part in numerous projects around Europe. She came up with and contributed to the creation of a unidirectional NMR relaxometer for non-invasive *in situ* analyses of porous materials.

Anna Laura Segre was the author of more than 250 scientific papers published in the most important international journals. She was invited to give lectures to the American Chemical Society Polymer Division. She acted as a referee for some international journals such as *Journal of Physical Chemistry*, *Langmuir*, and *Inorganica Chimica Acta*.

She acted as an evaluator on many committees within the CNR and was the head of science on many CNR projects. She was a member of the Scientific Committee of the Italian Group of Magnetic Resonance (Italian Chemical Society) and, in 1995, she was awarded a gold medal for her innovative studies in the magnetic resonance field. In 2002 she was the winner of the Sapio National Prize "Research 2002". She also acted as an evaluator for the European Community with regard to Food Safety.

Those who worked with her remember her vitality, her scientific curiosity, her innovative ideas and her constant hard work. She left her pupils and colleagues a legacy covering many varied areas of knowledge and she was a touchstone for everyone with her innovative ideas, her scientific curiosity, her hard work and above all her vitality.

Literature

Il CNR Ricorda Anna Laura Segre (The CNR remembers Anna Laura Segre) http://news.urp.cnr.it/varie/InRicordo DiSegre

Storia del CNR al Femminile (A Female history of the CNR) http://www.cnr.it/sitocnr/ IlCNR/Chisiamo/Storia/CNRalfemminile/ Segre.html

Ada Yonath (1939–)

Brigitte van Tiggelen

2009 was a remarkable year in the history of women's achievements: five out of the thirteen Nobel laureates were women. Among the laureates was the crystallographer Ada Yonath, 70. When she began with her research she knew that the topic was so fundamental that if a breakthrough could really be achieved, it would have Nobel-winning potential. But she also knew she was taking an exceptionally difficult topic and would remain alone on this path for a long time since it was considered difficult to the point of impossible.

Ada was born in 1939 in Jerusalem to Polish parents, who were Zionist and emigrated before her birth. Her father was a rabbi, and the owner of a grocery store he ran with his wife. Ada was passionately attracted to science from the start and would even perform experiments of her own at home. Her father died when she

Ada Yonath (credit: Micheline Pelletier).

European Women in Chemistry. Edited by Jan Apotheker and Livia Simon Sarkadi
Copyright © 2011 WILEY-VCH Verlag GmbH & Co. KGaA, Weinheim
ISBN 978-3-527-32956-4

was 10 years old and her mother moved to Tel-Aviv with her two daughters. Despite financial difficulties, since Ada got a generous fellowship, she allowed Ada to study in a very good High School. Returning from military service, Ada began studying chemistry, and got a master's degree in biophysics from the Hebrew University in Jerusalem and a PhD in X-ray crystallography at the famous Weizmann Institute in Rehovot. She held postdoctoral positions at MIT and Carnegie Mellon University and, as soon as she was back in Israel in 1970, she established at the Weizmann what was Israel's first protein crystallography laboratory. After returning from a sabbatical year at the University of Chicago, she also headed for 17 years the Max Planck Working Group on Ribosome Structure at the German Electron Synchrotron (DESY) in Hamburg, Germany, in parallel to her research activities in her homeland. She is Martin and Helen Kimmel Professor and leads the Helen and Milton A. Kimmelman Center for Biomolecular Structure and Assembly at the Weizmann Institute of Science. Helen Kimmelman, a New York benefactor has invested in her research since 1988. The US National Institutes of Health has also contributed to funding her work in Israel for more than 20 years.

The Nobel Prize in Chemistry 2009 was awarded to Ada Yonath, Thomas Steitz and Venkatraman Ramakrishnan for the detailed mapping of the ribosome, all the way to the atomic level. Any living body contains an impressive number (billions) of different proteins: they are the constituents for living tissues (e.g. collagen, the skin protein) and trigger or control the numerous chemical reactions required for life (hemoglobin carries oxygen from the lung to the muscles, insulin regulates the sugar level, trypsin digests the food). Though there is a huge diversity of proteins, they are all built from 20 different amino acids connected like pearls on a string by what is known as a peptide bond. The information on the sequence of the amino acids that build the proteins is contained in DNA, which is to be found in all cells. The genetic information is transported by messenger RNA to ribosomes, highly complex molecular complexes, which actually build the proteins. Ada Yonath wanted to reveal how the genetic code is translated into proteins, since the sequence of the amino acids in each protein is the key for its functional activity. In addition, she knew a great application could be for antibiotics, since half of the useful ones target the ribosomes, but she did not expect to be able to contribute to it. It was like climbing Mount Everest.

To achieve that goal, Ada Yonath wanted to establish the exact location of each and every one of the atoms of the ribosome using a very well-known technique: X-ray crystallography. But this requires the preparation of ribosomal crystals, suitable for diffraction experiments, which produce very complex diffraction patterns, showing how the hundreds of thousands of atoms are positioned in this huge molecular complex! The first step was thus to produce crystals of ribosome, which she achieved in 1980, using the ribosomes of very resilient micro-organisms, living under extreme conditions, such as those found in hot springs or in the Dead Sea. Yonath's assumption was that the ribosomes of these micro-organisms would be more stable, and would deteriorate less while being prepared, leading to a homogenous population with a higher chance of crystallizing. During the next 20 years, she patiently and tenaciously improved the crystallographic procedures, step by step. Eventually, 15 years after she started, other teams became convinced that the task was not that impossible and followed Ada Yonath's footsteps. Among others, one finds her Nobel co-laureates Thomas Steitz (1940–) and Venkatraman Ramakrishnan (1952–).

In August and September 2000 all three team leaders published the first crystal structures of ribosomes, with resolutions that allowed interpretation of the atomic locations.

What had started as an almost inaccessible quest ended up as a new prolific field, ribosomal crystallography, of which Ada Yonath is the pioneer and the founder. Moreover, she definitely paved the way to structure-based drug design of new antibiotics. By determining – in an incredibly short time – the structures of different complexes of antibiotics, she revealed the ribosome–antibiotics binding sites on the molecular level and provided precise insights into antibiotics selectivity and into the resistance acquired by pathogens to antibiotics. She elucidated the modes of action of over 20 different antibiotics targeting the ribosome. A better understanding of the mode of action of antibiotics could improve the existing drugs and lead to rational drug design to better target bacterial agents at the ribosomal level. Her work thus provides a mean of dealing with the crucial issue of drug activity and bacteria resistance to antibiotics, thus touching on a central problem in medicine.

In the beginning though things were tough, and she recalls being called "the village fool" for many years. However, it did not bother her: she had sufficient evidence (although hardly convincing anybody else) to remove part of her doubts about completing her scientific dream, even though at times she was not sure that it would work. Ada Yonath considers she was lucky, particularly in the beginning of her studies on ribosomes, as she met H.G. Wittmann (1927–1990), the director of Max Planck Institute for Molecular genetics in Berlin, who believed in ribosome crystallography, encouraged her, helped in establishing her research unit in Hamburg, and collaborated with her until he died in 1990. She also benefited from the

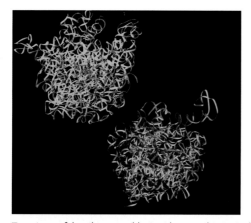

Two views of the ribosome (blue) with an antibiotic (red) bound to it (credit: Anat Bashan).

excellent research environment of the Weizmann Institute where she was allowed to pursue difficult research that was expected to yield groundbreaking results but could also fail. However, the contribution to humanity acknowledged through the Nobel Prize is mostly a consequence of her perseverance and consistent focus in the research agenda she had set at the beginning of her scientific career.

Ada Yonath is an outspoken person, and speaks her mind straightforwardly in all matters. Recently, she has expressed doubts about holding so many Palestinians in Israeli jails, believing this is counter-productive and that the motivation behind terrorism is a lack of hope. One of her cousins, Dr Ruchama Marton is an anti-occupation activist and founded "Physicians for Human Rights Israel", which received the 2010 Right Livelihood Award (Alternative Nobel). Ada was awarded the "Ettore Majorana-Erice-Science for Peace" prize in 2009: she has been one of the most prominent advocates of a major research facility being built in Jordan that will bring scientists from the Palestinian Authorities, Jordan, Israel, Egypt, Iran, Pakistan, Turkey and more surrounding nations to work together in harmony.

Ada Yonath is the recipient of numerous other prizes, among others, the Israel prize for Chemical research in 2002, the Louisa Gross Horwitz Prize of Columbia University, NYC in 2005, the Rothschild prize for Life Sciences in 2006, the Wolf Prize and the Paul Ehrlich-Ludwig Darmstaedter Prizes in 2007, a honorary doctorate from Oxford University in 2008, the Albert Einstein World Award of Science in 2008. Also in 2008 she won the L'Oréal–UNESCO Award for Women in Science. Though she never felt any discrimination as a woman scientist, she signed the L'Oréal–UNESCO Charter of Commitment for Women in Science, affirming along with other laureates her long-term dedication to promoting women in science careers. She recommends that young women and men go into science if they are really curious about a fundamental problem they want to solve. And there sure are many problems and challenges to face in our contemporary society. The accomplishments of Ada Yonath should confirm to the younger generation that their time has come, more than ever.

Literature

http://www.weizmann.ac.il/sb/faculty_pages/Yonath/home.html

http://www.weizmann.ac.il/sb/faculty_pages/Yonath/CV-AY.pdf

http://nobelprize.org/nobel_prizes/chemistry/laureates/2009/ in particular: MLA style: The Nobel Prize in Chemistry 2009 – Illustrated Presentation. Nobelprize.org. 4 September 2010.
http://nobelprize.org/nobel_prizes/chemistry/laureates/2009/illpres.html

Articles from the Jerusalem Post, www.jpost.com, especially:

Ada Yonath: Israel should release all terrorists, 10/10/2009

(http://www.jpost.com/Home/Article.aspx?id=157140)

Judy Siegel-Itzkovich, Former 'Village fool' takes the prize. Israeli scientist Prof. Ada Yonath and her chosen field of ribosomal crystallography have come out of the shadows into the limelight , 3/08/2008 – about the L'Oréal – UNESCO award for Women in Science
(http://www.jpost.com/Home/Article.aspx?id=94413)

http://www.mpg.de/english/illustrations Documentation/documentation/press Releases/2009/pressRelease200910081/index.html

Helga Rübsamen-Schaeff (1949–)

Susanne Bartel

■ Professor Dr. Helga Rübsamen-Schaeff was Senior Vice President and Global Head of Anti-Infective Research at Bayer HealthCare (Germany) and is now Chief Executive Officer of the Bayer Corporation spin-off AiCuris GmbH ('anti-infective cures') which she has headed since its foundation. Her research interest as a bio-chemist was in cancer research where she pioneered the study of retro-viruses. Later, she began research on Aids in which she developed tests and proved the existence of several different HIV viruses. Within Bayer AG and later AiCuris GmbH she has developed new drugs. One of the many awards that Professor Rübsamen-Schaeff has gained is the Mestermacher prize 'Manager of the Year 2004'.

Professor Dr. Helga Rübsamen-Schaeff says: "Women have to understand that natural science is not a closed book".

Her point of view is that "natural science is a lot of fun". Especially, the field of chemistry offers a huge variety of possibilities for one's professional life, and, being a highly successful scientist for more than 25 years, Professor Rübsamen-Schaeff knows what she is talking about.

However, when she started to study chemistry at the University of Münster the number of female students in this area was extremely low, 10% female students compared with 90% male students. In the late 1960s young women who enrolled in universities, were suspected of searching for husbands rather than being really interested in their studies. Chemistry wasn't Prof. Rübsamen-Schaeff's first choice. Initially, she was more interested in medical science but, at the age of 18, she was too afraid of making mistakes if working as a GP. "I would never have forgiven myself had I made a mistake", she says. Influenced by her mother, who was always very interested in biological and environmental links, she chose to give chemistry a try. "It was the subject I had the least knowledge of", Prof. Rübsamen-Schaeff admits with a laugh. In addition the skills gained in school were not pre-requisites for this course of study. Her girls' school's main focus lay in philosophy, art, and extinct languages such as Latin. Chemistry was considered of minor interest only and consequently taught for not much longer than a year. Even if approval for female chemistry students was not expressed by her fellow students she

European Women in Chemistry. Edited by Jan Apotheker and Livia Simon Sarkadi
Copyright © 2011 WILEY-VCH Verlag GmbH & Co. KGaA, Weinheim
ISBN 978-3-527-32956-4

Helga Rübsamen-Schaeff

was able to find friends who helped her eliminate the gaps in her education. After having successfully passed the pre-examen, Rübsamen-Schaeff was granted a scholarship by the Studienstiftung des Deutschen Volkes (German National Academic Foundation).

Nonetheless, she still doubted whether chemistry was really the right choice or if she should take a different course. She then came into contact with the development of cancer cells and their way of working in a summer school seminar. Subsequently, the young woman decided to specialize in the area of biochemistry. "My purpose lay before me, from then on I knew what I wanted to do", Rübsamen-Schaeff states.

Already at the age of 24 she had been awarded PhD 'summa cum laude'. Her work as a post-doctoral researcher then led her overseas to work at Cornell University, Ithaca (NY), one of the Ivy League universities in the USA. During that period of time she deepened her knowledge in the field of biochemistry. Even though Rübsamen-Schaeff had achieved a lot at an early age her work had not really been recognized before she went to the US. She states that working conditions for women were completely different there compared to her homeland. "In Germany nobody hindered me but I hadn't really got a good reputation", Rübsamen-Schaeff remembers. Back in Germany she concentrated fully on cancer research by looking for the perfect system to study the development of healthy cells into carcinogenic ones. The Rous-Sarkoma-virus with its one and only gene proved to be that ideal model, and during the following years the scientist directed her attention to researching that virus. Today the study of retro-virus (RNA virus) has become a standard in renowned cancer research but Rübsamen-Schaeff pioneered the method.

While still fully absorbed in the field of cancer research in the 1980s, Rübsamen-Schaeff became involved in the area of Aids research. Because of the diverse symptoms of the illness she was sure that Aids was caused by more than one virus. Her intuition proved right and together with her colleagues she discovered various HIV cultures. At that time she had been awarded the qualification of a university lecturer in the field of biochemistry and worked at the Chemotherapeutic Research Institute Georg-Speyer-Haus in Frankfurt. It was the Managing Director, Professor Brede, who approached Rübsamen-Schaeff to become Scientific and Managing Director of the Institute. She admits to having been unsure if she would be able to handle this position. Prof. Brede, however, had no doubts and simply told her "to just try it". From her point of view her doubts seem to be a typically feminine reaction. But in the course of her career Rübsamen-Schaeff learned that fathers of daughters, like Prof. Brede, are more willing to give ambitious and well-trained young women a chance.

Within a few years of becoming the new Director she managed to transform the Georg-Speyer-Haus from a low-key institution with an annual budget of 20 000 Deutsche Marks into an excellent research institute with a budget now of more than 100 000 Deutsche Marks per annum. Rübsamen-Schaeff designed HIV tests and applied for patents on these. In addition she worked on the development of treatments against the virus and cooperated with Hoechst and Bayer.

At the same time as she started working as Managing Director, Rübsamen-Schaeff became the mother of a son. But motherhood did not prevent the ambitious researcher from working. As Rübsamen-Schaeff says, her child was a planned child. With the help of a nanny and her husband she was capable of dealing with both positions. An understanding and helpful partner, as well as relatives, are the source of a well-organized and successful family life, but parents had and have to be willing to manage the family in such a way that it is no hindrance to a business career. Rübsamen-Schaeff feels it necessary for parents to be prepared in case governmental support in child day care is lacking.

In 1994 Prof. Rübsamen-Schaeff moved into industry. Bayer AG offered her the position of Director of the Antiviral Research Department. She now had the opportunity not only to do research but to also develop new drugs. With that she entered a new world. On the one hand she experienced the long time scale of launching a new drug, from development to actual application. On the other hand, she then had to deal with an annual budget of about 17.5 million Euros as compared to a few hundred-thousand at the Georg-Speyer-Haus. Seven years later, in 2001, she accepted the position of Senior Vice President and Global Head of Anti-Infective Research at Bayer HealthCare (Germany). When the Bayer Corporation intended to spin off the department to an independent corporation, again it was Prof. Rübsamen-Schaeff who was asked to become Chief Executive Officer.

"This decision gave me nightmares", she states. "It was a difficult time in my career". She took the new step bravely, although it was not easy to leave colleagues and the corporation behind. Furthermore, being Chief Executive Officer of the newly founded AiCuris GmbH (the name is derived from 'anti-infective cures') also meant re-positioning her business life. "It was very complicated to create perfect conditions for my colleagues and the new corporation while still working at the old corporation".

Rübsamen-Schaeff sees her success arising from her willingness to take risks, following her own inclinations, hard work and leaving the old paths. Women can be very successful if they are prepared to do this. Furthermore, the lack of experts in natural sciences offers good opportunities to young women. However, Rüb-samen-Schaeff also admits that the basic conditions are highly important. This includes, for example, well-organized child day care as well as society's willingness to accept successful working women. And women themselves have to take the necessary steps. In her eyes children do not prevent one from having a career. However, she considers long breaks between the birth of a child and the return to work as "poisonous". It is important that women keep in touch with their work place during their maternity leave and return to work as soon as possible.

The successful academic who received the 'Bundesverdienstkreuz 1. Klasse" of the German State (Federal Cross of Merit 1. Class) and the Mestermacher prize 'Manager of the Year 2004' still has dreams for her own career. "I wish that one day one of the drugs we invented at AiCuris will be sold in pharmacies all around the world".

Katharina Landfester (1969–)

Katharina Al-Shamery

■ Professor Dr. Katharina Landfester is the first female director of a Max-Planck research institute in chemistry (second female director of 128 MINT related institutes) who was appointed in 2008 as one of the directors at the Institute for Polymer Research in Mainz. She is a polymer chemist and has pioneered the use of
mini-emulsions to make new materials with broad applications ranging from
heterogeneous catalysis to drug delivery.

> *"Yes, you can, if you have convinced yourself that you can do it",*
> *would be the motto of Katharina Landfester.*

Katharina Landfester was born, as the first child of three, in 1969 in Bochum, a
large city in the Ruhr valley in Germany where she lived until the age of 12. Her father was a university professor of Ancient Greek. He was the first important person in Landfester's life and he never gave up encouraging her at every stage of her
career. Her mother worked in Slavistics for ten years until she decided to become
an artist who also gave drawing classes for children.

The family moved to Gießen, a town in central Germany when her father got a
new appointment at the local university. Gießen is the city of Justus von Liebig with
one of the ten most important museums in chemistry, including the original laboratory and lecture hall of von Liebig. So maybe it was natural that Landfester became interested in chemistry when she was in the 9th grade. "Chemistry is not for
women" was the comment she heard when she tried to communicate her future
plans. So she first changed her mind and thought about becoming a teacher in
Latin, Greek and history. However, one day before the baccalaureate she decided
that she was not willing to adapt herself to the ideas of the surrounding society and
was dedicated to proving to the world that she would succeed in chemistry.

Maybe as an act of stubbornness she decided to specialize in technical chemistry
and so she chose the Technical University of Darmstadt. When she started to study
female students made up about 20% of the freshmen (in chemistry), which was
possibly because it was a technical university. "Look left and right at your female
colleagues, they will all get married soon and stop their studies" was the first thing

European Women in Chemistry. Edited by Jan Apotheker and Livia Simon Sarkadi
Copyright © 2011 WILEY-VCH Verlag GmbH & Co. KGaA, Weinheim
ISBN 978-3-527-32956-4

Katharina Landfester (photograph provided by the author).

she heard in the lecture hall. Landfester proved them wrong. She hated comments like "May I help you?" when she could sense the underlying thought "because you are a woman and therefore cannot do this". For her this was a reason to double her efforts to succeed on her own. For the experimental part of her diploma thesis she moved to the Ecole d'Application des Hautes Polymères in Strasbourg where she worked with Professor M. Lambla for twelve months, including a prolonged stay after her final examination. Though she had studied French at school as a fourth language she took intensive courses to be able to communicate better. In France she was impressed by the social system that allowed women to go back to work soon after childbirth as childcare was easily available.

For her PhD thesis she decided to go back to Germany to the Johannes Gutenberg University of Mainz where she got her doctoral degree in physical chemistry in 1995. Her work with Professor H.W. Spiess at the Max Planck Institute for Polymer Research included the synthesis and characterization of core–shell latexes which she characterized with transmission electron microscopy and solid state NMR. Spiess became an important mentor who supported Landfester's career. After spending another year as a group leader at the Institute Landfester decided to go to the Lehigh University Bethlehem (Pennsylvania) in the USA, to Professor M. El-Aasser as a postdoctoral fellow. This step turned out to be a key milestone as she came into contact with the mini-emulsion technique. The method was not yet elaborate at that time but Landfester immediately recognized its enormous potential.

During her stay she enjoyed going to concerts in Philadelphia and the (Metropolitan) opera in New York. She was particularly impressed by Diane Wittry, a famous female conductor whom she loved to watch when Wittry directed a huge orchestra. Wittry started to be one of her female heroines beside Jutta Limbach, President of the German Federal Constitutional Court.

Landfester moved back to Germany in 1998 where she started to further explore the mini-emulsion technique within the group of Professor M. Antonietti at the Max Planck Institute of Colloids and Interfaces in Golm with a Liebig stipend of the Fonds der chemischen Industrie (FCI). Antonietti became the second person to support her professional career. Her focus was now on new possibilities for the synthesis of complex nanoparticles.

In 2000 Landfester met her future husband, a medical doctor. The year 2001 was a particularly successful one for her as she was awarded the Reimund Stadler price of the Gesellschaft Deutscher Chemiker (GdCh) and Dr. Hermann Schnell Stiftung prize. In 2002, Landfester got her habilitation in physical chemistry at the University of Potsdam. Soon after, she became a member of the young academy of the Berlin-Brandenburg Academy of Sciences and the German Leopoldina from 2002–2007 for which she acted as a spokesperson in 2003/2004.

Shortly after her habilitation, in 2003, she accepted a chair (C4) of macromolecular chemistry at the University of Ulm. Though her husband was not very keen to move from Golm (which is close to Berlin) back to Ulm, his birth place on the border with Bavaria, he supported his wife. He was aware what he was up against as she made clear right from the beginning of their relationship that she would not compromise. Landfester convinced the university to help get a position for her husband though the dual career problem was not yet an issue at that time in Germany.

In Ulm, close to the clinics and the large medical faculty, she started her activities in the field of biomedical applications of mini-emulsions in cooperation with several medical groups. Together they looked at the interaction of nanoparticles with different cell compartments, the labelling of cells and the delivery of substances to specific sites. In 2006 Landfester gave birth to her first child, a daughter. Two weeks later she was back at work, however, always accompanied by her daughter for the first year. Certainly taking her daughter to meetings with the rector helped to accelerate the building of a house for child care. When she had to lecture, her husband took care of the baby or she simply took her daughter into the lecture hall. Being a mother inspired her to establish a laboratory, called the EMU (emulsions and macromolecules in Ulm) laboratory, for kindergartens and schools, where groups can perform experiments every week on emulsions, milk, soaps, polymers, recycling, and so on.

Finally, in 2008, she joined the Max Planck Society as one of the directors of the Max Planck Institute for Polymer Research in Mainz and thus was the first female director in chemistry in the Max Planck Society and the second female director in a MINT subject overall. In 2009 her second daughter was born and again is still with her all the time.

"Authority as a result of competence" is the answer when Landfester is questioned about her management style. Though professionally she has achieved mainly everything possible in Germany she is still driven to broaden her research work and to make her field even more visible than it already is.

Yes, she can...

Index

European Women in Chemistry. Edited by Jan Apotheker and Livia Simon Sarkadi
Copyright © 2011 WILEY-VCH Verlag GmbH & Co. KGaA, Weinheim
ISBN 978-3-527-32956-4